MVFOL

Golden Highlights Library

Clocks & Watches

E. J. Tyler

Golden Press

Published in 1974 by **Golden Press, New York,**
a division of Western Publishing Company Inc.
Library of Congress Catalog Card Number: 73–87966

Created, designed and produced for
Western Publishing Company, Inc. by
Trewin Copplestone Publishing Ltd, London

Printed in Italy by
Officine Grafiche Arnoldo Mondadori, Verona
Filmset by Photoprint Plates Ltd, Rayleigh, Essex
World rights reserved by
Western Publishing Company, Inc.
GOLDEN and GOLDEN PRESS ® are trademarks
of Western Publishing Company, Inc.
ISBN: 0 307 43118 5

Contents

Page 1 *Column sundial, French, 17th century. The dragon is rotated until its tail is opposite the appropriate sign of the Zodiac.*

Acknowledgments

The photographs for this book were taken by Geremy Butler except those credited below to: Hamlyn Group Picture Library: 44; Bernard Mason: 22; Oeffentliche Kunstsammlung Basel, Switzerland: 6l; Kenneth D. Roberts 71t, b; Schatz Rombach (Successors) Limited: 74; Synchron Limited: 80b. The publishers also gratefully acknowledge: The Clock Shop, London: 40t, b, 60, 62, 66t, b; The Haberdashers Company: 24; Held, West Germany: 19r; Mr B. Hutchinson: 58t, b, 59, 68, 69, 72, 73; Ingersoll Limited: 76l, r, 77; Kingston Antiques, Surrey: 32l, r, 33, 52l, r, 61t, b; E. Pitcher & Company: 56l, r, 57l, r; Science Museum, London: 3, 4b, 5, 6r, 7, 8l, r, 10t, b, 12, 13t, b, 14, 15t, 16, 17, 28, 19l, 20, 21l, r, 28, 29l, r, 30, 31, 34l, r; Rev. Graham Smith: 42, 43; Strike One, London: 23, 25, 26, 32t, b, 38t, b, 39, 53; Wellington Museum, London: 4t.

Dover Castle turret clock, probably early 17th century. The frame and wheels are wrought iron as is the foliot balance seen at the top.

Introduction

ALL time measuring devices have one thing in common: they perform some action repeatedly and the number of occasions on which this action is performed after a given instant measures the time which has elapsed. The sun crosses the sky once a day; the water clock may last for several hours before it has to be reset; the sand glass acts more rapidly than this, varying from an hour for the sermon glass to three minutes for the egg boiler and still less for the navigator. With the mechanical clock, the rapidity is increased still more. The balance of the clock preserved in Salisbury Cathedral (1386) beats about once in four seconds, the iron Gothic wall clocks about once in $1\frac{1}{2}$–2 seconds. The introduction of the pendulum speeded up the rate of time measurement. The earliest pendulum clocks beat about 3 times to the second, but when the long case clock was established, pendulums beating once per second became popular. Mantel clocks, of course, had much shorter pendulums and therefore beat faster.

The early watches beat at a slightly faster rate than twice per second, but when the balance spring was introduced, four or five times to the second became usual (14,400 or 18,000 beats per hour). Modern practice is improving on this: 6 times to the second (21,000 per hour), and even as high as 10 to the second (36,000 per hour). The limit of mechanical vibration is at present in the Bulova Accutron which has a tuning fork vibrating at 360 times per second (1,296,000 per hour).

The latest developments in time measurement use the vibrations of a quartz crystal instead of a mechanical body. The number of vibrations chosen may be 16,384 or 32,768 per second, and very much higher numbers than this are being planned. A quartz crystal timekeeper gives an accuracy of within one minute per year, and work is still being done to improve on this. The history of timekeeping through the centuries has consisted of the speeding up of the vibrating unit to achieve ever greater accuracy.

During the period in which timekeepers have been made, certain types that have become more well known than others–for good performance, bad performance, cheapness, high price, and so on. This book deals with a selection of some well known types. However well or badly they function, or however rapidly or slowly they beat, each in its own way forms an integral part of the history of time measurement.

The Turret Clock

THE earliest mechanical clocks were made about 1300 and, as the technology of the time was founded on the blacksmiths' craft, they were produced in wrought iron and were necessarily large. They were mainly used by the church, and it is probable, although not certain, that many of them were originally located at ground level and only later were to be found in elevated positions in towers. Their function was to strike on a bell or bells at intervals, and the idea of recording the time by means of a dial only came later.

The turret clock now preserved in Salisbury Cathedral is claimed to be the oldest in Britain (1386). It is controlled by a foliot balance that takes about 4 seconds to make one vibration.

An elaborate chamber clock suffers from the attentions of the household pet. Detail from The Dissolute Household *by Jan Steen (1626–79).*

The early clocks were made with their trains of wheels placed end to end, but after the introduction of the pendulum in clock work in the 17th century, clocks began to be made with the trains side by side. The frame was made more solidly and the vertical bars for carrying the pivots of the arbors were fastened with nuts instead of wedges. The movement fitted compactly against the wall of the tower and was usually boxed in.

The building of the new Houses of Parliament in the mid 19th century raised the question of what type of clock should be fitted in the tower. The clock was eventually designed by E. B. Denison and differed in two ways from the type of turret clock known at the time: the frame was made horizontal and arranged so that any wheel could be removed without disturbing the others. The clock was also fitted with Denison's gravity escapement giving a constant impulse to the pendulum. A smaller clock made to try out these principles is now in the church at Cranbrook, Kent.

The Chamber Clock

THE original turret clock was intended to serve a community and was used to tell the time by sound over a wide area, before dials came into use. Smaller clocks for the home could only be made after the idea of recording the passing of time by means of a dial had become popular and the necessary technology had been developed. This technology was provided by the locksmiths, who worked with the metal cold and cut it by means of files, in contrast to the blacksmiths who shaped the hot metal by means of hammers, only using a file for finishing.

The earliest domestic clocks were made of iron, like the turret clocks, and were generally based on the same design. The trains of wheels were arranged one behind the other and the frame of the clock suggested the tower in which the larger clocks were usually housed. It was necessary to add extra wheels to the train in order to limit the fall of the weights, for otherwise the clock would have had to be hung too high to be readily visible. Some clocks were made to sound an alarm at regular intervals. This attracted the attention of a watchman, who would then ring a large bell in a church or city tower to tell the time to the inhabitants at night. Such clocks are still to be seen in museums in Germany and

Chamber clock movement with wide spaces between the hand-filed teeth and decorated corner posts.

Chamber clock, made by Ulrich and Andreas Liechti, Winterhur, Switzerland, 1596–a scaled down version of the turret clock of the period.

their making led to the use of alarm work on domestic clocks in addition to the hour striking. Alarms were usually operated by inserting a pin into a hole in the wheel carrying the hour hand, but they could only sound at fixed times such as the whole or half hour. Alarm work that could sound at any desired time was developed later.

The oldest wall clock in the world is claimed to be that in the Mainfrankisches Museum in Wurzburg, Germany (suggested date 1352). This clock is of the type used by a watchman, but there is no evidence to prove that the watchman type was made earlier than the domestic striking clock.

The iron chamber clock is usually associated with South Germany and Switzerland and, to a lesser extent, with France and Italy. The type was produced until about the middle of the 17th century, but after the invention of the pendulum, wealthier people wanted something more sophisticated and the chamber clock ceased to be made in its original form. It belonged, after all, to the pre-pendulum era, being controlled by a wheel balance or by a foliot as were the early turret clocks. The makers of these clocks remained faithful to iron as a material, although brass had been introduced into horology in the 16th century.

Fortunately, many examples of iron chamber clocks have been preserved and can be seen in museums throughout Europe.

The Nocturnal

THE nocturnal is the night equivalent of the sun dial. Its action is based on the fact that the Pole Star appears to remain stationary in the heavens while all the other stars make one revolution round it during a sidereal day. For the purpose of observation, the two constellations nearest to it are chosen: the Great Bear, sometimes known as the Plough or the Dipper, and the Little Bear. If the sidereal day and the mean solar day corresponded, the time could be read from the sky by regarding the Pole Star as the centre of a clock dial and the Great Bear or Little Bear as the hour hand, but it is because of the difference between the two types of day that an instrument is necessary to adjust sidereal time to mean time.

The nocturnal itself consists of two movable discs fastened to the base of the instrument at the center, each bearing the scale of hours 1–24. Each disc is marked with an L or G according to whether it is to be used with the Little Bear or the Great Bear. The indicator is set to the date, which is marked on the edge of the baseplate, then the Pole Star is observed through the hole in the center. A long arm mounted on the same center is then brought round until its edge coincides with the appropriate stars. The time can then be read on the scale where the edge of the arm crosses it.

Far left *Chamber clock in a domestic setting–the home of Sir Thomas More. It is hung high to give maximum fall to the weights. Drawing by Hans Holbein (1497/8–1543).*

Center *Nocturnal. The instrument is held with the handle downwards and the Pole Star viewed through the center hole. The dates round the edge also show the Zodiacal signs.*

Left *Rear view of the nocturnal. The scale on the left allows angles of elevation to be measured and other astronomical information is shown in the center.*

Nocturnals were first made about 1520 and were fairly popular during the 17th and 18th centuries. They were much simpler to make than watches, and in the days before street lighting the stars were much easier to observe than they are today. Both metal and wooden examples are found and sometimes the disc with the hour markings is provided with teeth so that the time can be felt with the fingertips without the need for artificial light.

The Portable Sundial

THE sun was the first natural timekeeper. Primitive man would have noticed how the shadows cast by it moved roung during the day and how they were longer in winter than in summer, when the sun did not rise as high above the horizon.

Any fixed object could be used as a primitive sundial by making marks on the ground near it corresponding to various periods of the day which were reached and passed by the moving shadow. However, when a man was away from his home he needed some kind of portable instrument to tell the time by the sun. Two kinds of portable sundial are known, one which measured the time by the sun's altitude, the other by its position in the sky. The first was the older type and an example of one of these made in the 10th century has been excavated at Canterbury. The principle of the instrument was that the style or gnomon which cast the shadow pointed at the sun, and according to the altitude of the latter a longer or shorter shadow was cast on the scale. The instrument first had to be adjusted to the time of year and to the time of day–before or after noon.

A popular form of portable sundial working on this principle was the pillar dial, which has been used until recently in the Pyrenees. The gnomon was at the top fastened to the upper part of the pillar which was free to rotate. The gnomon was turned to that part of the circumference of the pillar appropriate to the time of the year, and the end of its shadow would point to the time on the scale at that point. The material used for these instruments was wood, metal or ivory.

Another form of portable dial based on the sun's altitude was the ring dial. While the pillar dial was only true for the latitude for which it was designed, this instrument was adjustable. The ring containing the hour markings had its angle adjusted to the latitude and the sun shone through a small hole in a slide that could be adjusted to the time of year according to the sign of the zodiac. The instrument, held by its supporting ring, was turned until the point of light shining through the hole rested on the hour scale, giving the time.

For a dial to tell the time by the position of the sun in the sky it was necessary to know the compass direction, and many of the early dials included a compass. The two halves of the instrument opened like a book and the thread joining them acted as a gnomon. Sometimes small styles told the hours by the Italian system, on additional dials.

Portable dials were in use until the 18th century, but the Industrial Revolution brought with it a need for more accurate timekeeping, and soon these dials were superseded by watches. A few years ago, compasses were sold with a dial plate and gnomon carried by the needle, for telling the time by the sun, but these instruments were only intended as a novelty and not for serious use. In countries such as Britain, where the sun is frequently not visible, a sundial is of less use than in countries with a more settled climate. While British dial makers were known, it is possible that many dials used there were made overseas, for records have been preserved of bone, ivory and wood dials imported through Rye, Sussex in the 16th century.

Pocket sundial of German make, early 17th century. The gnomon is a string which can be adjusted for six latitudes. Vertical and horizontal styles allow for other time systems.

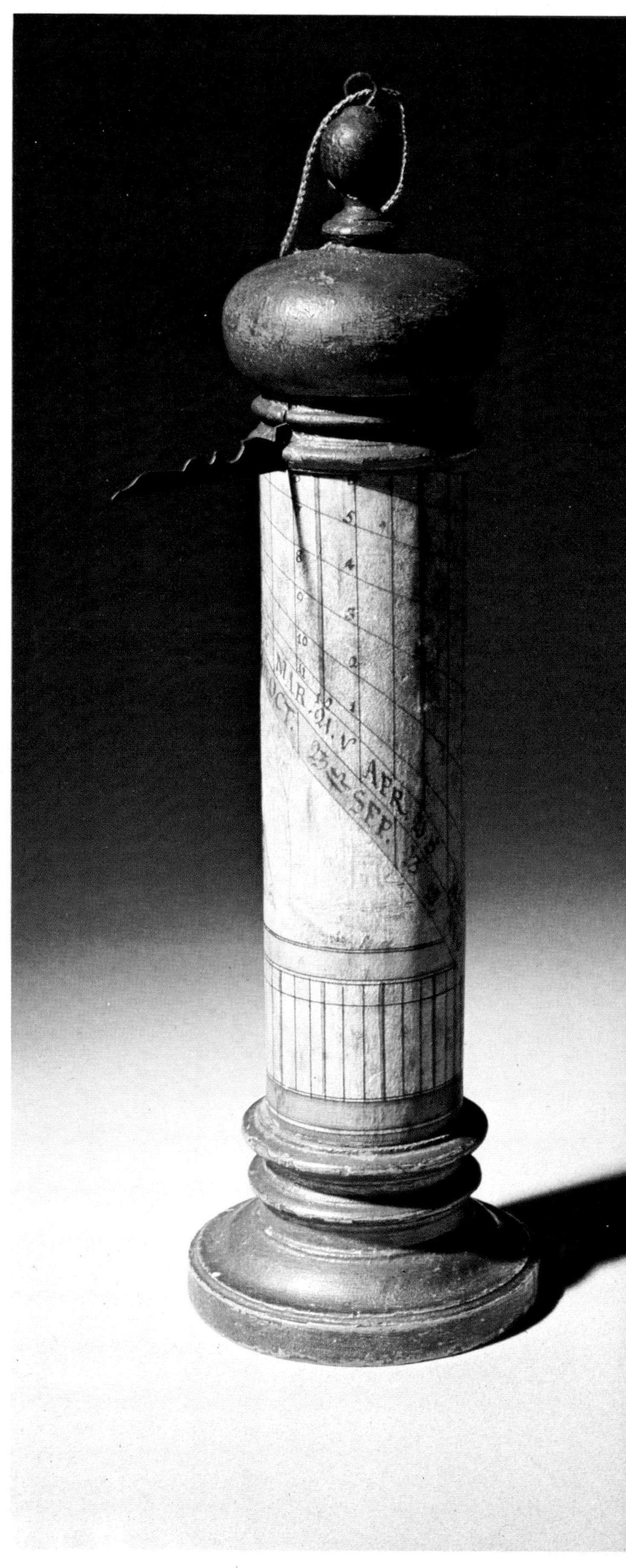

Wooden pillar dial, 17th century. The top is rotated to bring the gnomon above the appropriate season.

Ring dial, modern. The inner ring bearing the hour scale is adjusted to the latitude. The sun shines through a hole in the center jewel.

The Stackfreed Watch

THE mainspring as motive power for timekeepers was introduced at some time during the 15th century. The verge escapement, which was the only form known at that time, was very sensitive to changes in the power applied to the train, and as a spring exerted less force as it ran down, the clock or watch went more slowly as the time for winding approached. To make matters more difficult, the spring did not lose power steadily, but lost it rapidly at first, more slowly during most of the running period, then rapidly again at the end.

The earliest attempt to overcome this variation in power was by means of the fusee, a piece of metal shaped roughly like a cone with a spiral groove cut round it, connected to the drum or barrel containing the spring by means of a gut line. The fusee was rotated to wind the

clock, and the line was pulled off the barrel, thereby winding the spring. At the beginning of the going period, the line pulled on a small diameter of the fusee and as the spring unwound and became weaker, the line pulled on an increasing diameter, so that the power was made nearly even throughout.

An alternative method devised to equalize the pull of the spring was the stackfreed. The origin of the word is obscure. By mixing German and Dutch it could be translated as "stop spring", but as all the clocks and watches fitted with the device appear to be German, it is strange that a Dutch element should have crept into the word.

The stackfreed is an interesting example of the early use of a cam. The cam was carried on a wheel resting on the top plate of the watch and meshed with a pinion on the mainspring arbor. The wheel was not completely furnished with teeth, a space being left uncut, and this acted as a stopwork, preventing the spring from being completely wound up or run down. The cam had an indentation and when the watch was fully wound, a roller on a strong spring rested in the indentation and as the watch began to run, the roller was forced to climb out of the indentation and so resist the power of the mainspring. By the time it had climbed out completely, the first power of the mainspring had gone and further rotation of the cam had little effect, but when the going period was nearly over, the indentation had come round again and the roller then ran down the slope, thus adding to the power of the main-spring and making up for the falling off in power at the end of the run.

The idea was ingenious, but not precise enough to give good timekeeping. Some stackfreed watches were provided with a hog's bristle to interrupt the movement of the balance and attempt to get a steadier rate, but the whole arrangement is too uncertain for accurate timekeeping. Occasionally variations on the usual shape are met with that resemble a kite shape and the roller is replaced by a shoe.

Opposite top *Stackfreed watch movement showing the roller and cam, the stopwork to prevent overwinding and the hog's bristle regulator acting on the arms of the balance.*

Opposite bottom *Stackfreed watch, late 16th century. Note the use of Roman figures I to XII and Arabic figures 13 to 24.*

All the stackfreed watches known are suspected to be German in origin. They seem to have been made from the mid 16th until the early 17th century, but with the advent of more finely made watches at this time the stackfreed would have been an anachronism and it died a natural death. Although it was a clumsy device, it showed ingenuity and must have been highly thought of, to have been made over such a long period. It was simpler than the fusee and allowed the watchmaker a little more space between the plates of the watch. It is a pity that we shall probably never know the name of its inventor.

The Nuremberg Egg

TRADITION has it that the watch was invented about 1510 by a locksmith of Nuremberg named Peter Henlein. The authorities of the city have set up a statue of him, but it is not certain that he was the inventor. Other cities apart from Nuremberg may have been producing small portable timekeepers during this period and Italy and France might well have possessed craftsmen of sufficient skill to produce this device. Nuremberg has, however, acquired the reputation of being the birthplace of the first watches, traditionally oval in shape and known as "Nuremberg eggs".

The earliest watches known today are all spherical or drum shaped and comparatively rough with movements made of iron. Oval watches are to be seen in many museums but appear to date from the early 17th century, by which time movements had become more refined and were being made mainly of brass. The watches, although oval in outline, were often flattened on the front and back and therefore were less egg shaped. By this time watches were being made all over Europe and were no longer particularly associated with Nuremberg. The confusion over the name may have arisen because an early writer described the Nuremberg watches as *Uhrlein* (little clock) and this was misread as *Eierlein* (little egg).

The heyday of the oval watch was probably about 1610–1625, for the design was then most popular among fashionable women. A number of English oval watches have been preserved and one in particular is well known, having been made by John Midnall in Fleet Street about 1625 and used by Oliver Cromwell. This watch provides an early example of a dial protected by glass and indicates the beginning of

the custom of wearing watches in the pocket. Before this, they were worn on a chain round the neck, and were decorated on the outside. The custom of wearing watches in the pocket is attributed to the Puritans, who disliked outward show. A watch for the pocket needs to have a smooth exterior if it is to be capable of being easily withdrawn and replaced.

A number of French oval watches are also known. Some of them have cases decorated with enamel instead of the usual engraving, and while some have a smooth case exterior, others possess the older type of case intended to be hung from the neck. Some of these French watches are fitted with oval glasses; several French design books are known showing designs for such oval watch cases. Some of the preserved examples are credited with a date earlier or later than 1610–1625.

There is no particular advantage in making a watch oval, and the taste for oval watches can be attributed to fashion. The circular watch has been made for centuries and it is only at the present time that it is being seriously challenged by other shapes. For practical considerations the circular shape is generally regarded as the best.

Opposite *English oval watch by Richard Jackson, London, about 1630. Note the stumpy figures and decorated hand.*

Above *The watch movement showing the decorated balance cock, the ratchet for setting up the mainspring and the engraving on the edge of the plate.*

Left *The opened watch showing the fusee and gut line on the barrel. The top and bottom of the case are fairly flat, thus making the shape less like an egg.*

Early lantern clock made by Thomas Knifton, about 1650. The clock has a single hand, a narrow chapter ring and a wheel balance.

English lantern clock movement by John Drury, London, about 1760. The clock is fitted with spurs to keep it away from the wall.

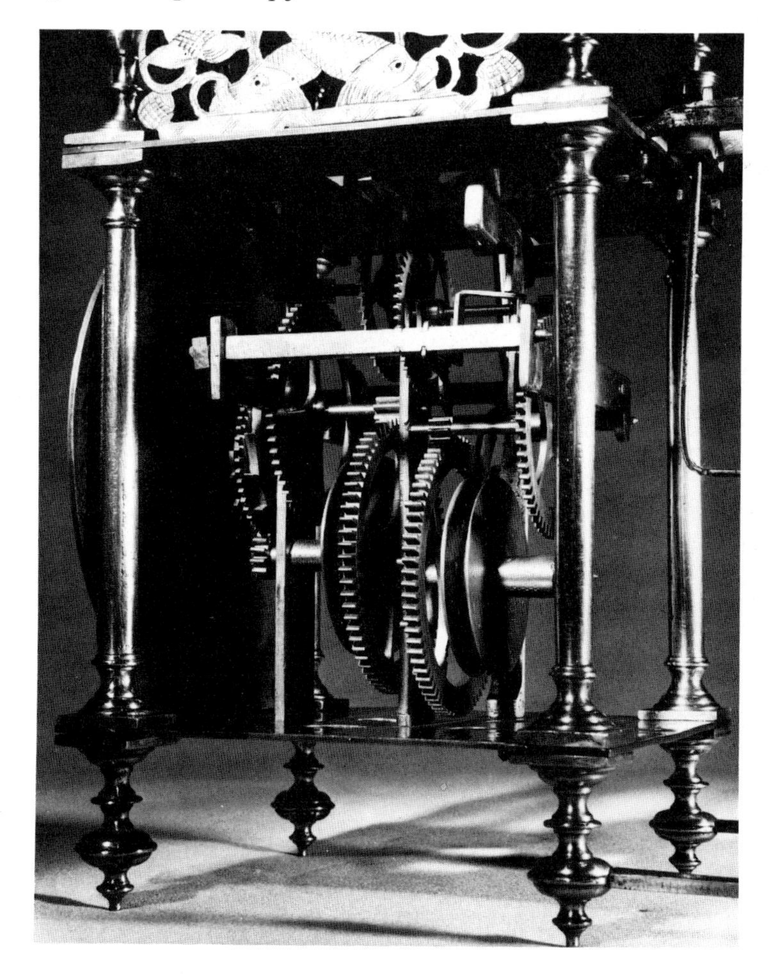

The Lantern Clock

UP to the end of the Elizabethan era there were scarcely any English clockmakers. Practically all the clocks and watches used in Britain were made abroad or by foreign workmen who had settled in the country, and bore the designs of the countries from which their makers had originated–mainly French styles by French workmen.

At the beginning of the 17th century the first truly English style began to emerge. It was a weight clock based on the Gothic wall clocks found in Europe, but included a large proportion of brass in the movement, while the Gothic clocks were made of iron, and the external finish consisted of brass instead of sheet iron decorated with paint. The type is generally known as the "Lantern" clock, but the derivation of the name is obscure. The shape of the clock may be said to resemble a lantern, but the probable explanation is that with the prevailing French influence in horology at the time the clocks were called *laiton* (brass) clocks to distinguish them from the earlier iron ones, and the English mispronounced *laiton* as "Lantern". The names "bed-post" and "Cromwellian" have also been applied to the type.

The clocks tended to be shorter than the Gothic wall clocks and were originally fitted with wheel balances as the brass frets would make a foliot inaccessible for adjustment purposes. The clocks were regulated by a depression in the top of the weight of the going train containing lead shot that could be removed or added to, in order to adjust the rate at which the clock ran. The earliest dials had quite narrow chapter rings with stumpy figures and a single hand of iron with a tail for assisting the adjustment of the hand to correct time. Striking was always fitted; the types with alarm work only evolved later in the 17th century.

After the application of the pendulum to clockwork in 1657, lantern clocks were also made with pendulums, and as well as new clocks being fitted with the device, older clocks were also converted. The earliest type of pendulum was short and was applied to a verge escapement with the scapewheel on top of the clock. After the invention of the anchor escapement, long pendulums were applied. The clock would be hung on the wall by means of an iron loop, with spurs at the rear to keep it in position, and the pendulum would swing between the spurs. Later clocks had an endless

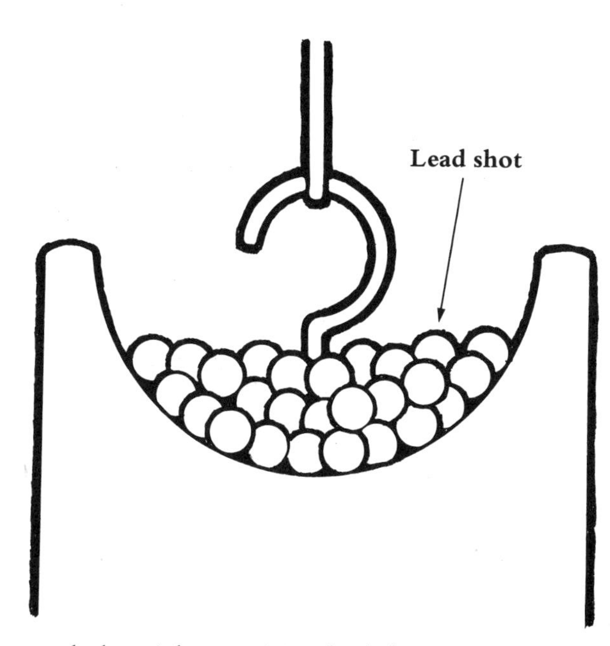

Lantern clock weight containing lead shot.

rope or chain allowing one weight to drive both going and striking, and earlier clocks were converted to this method. As the early clocks were arranged to have their great wheels rotating in opposite directions, the winding gear had to be modified when conversion was carried out and the striking train suitably adapted.

There are now very few old clocks in their original state, but when a clock has been converted, traces of the original mechanism are usually apparent. Many of the clocks are now being re-converted to balance control. During the Victorian era many of these clocks were fitted with contemporary spring driven movements and adapted for use on a mantelshelf but while the performance of the clock was improved by this, the clock was ruined as a work of art. The Victorian era also saw many of these clocks sent to the scrap heap. Even a few years ago the type was not highly appreciated by collectors, but today there is much competition to acquire them.

Detail of Swiss clock movement, showing how soundly it is made, despite its primitive design.

The Swiss Wooden Clock

THE clocks of the Black Forest have achieved fame in many parts of the world, but the industry there did not develop fully until the middle of the 18th century. It is not generally known that quite elaborate wooden wheeled clocks were being produced in Switzerland a century earlier than this.

The Swiss clocks were derived from the Gothic iron chamber clocks, were driven by weights and controlled by a wheel balance made of wood which was surrounded by a narrow metal band to overcome the tendency to split. As the weights were hung on cords which gripped their sprockets merely by catching in small projections, counterpoise weights were necessary to keep the opposite ends of the cords taut. The frames of the clocks were also made of wood and contained brass bushes for the pivots which were formed of wire. The arbors were decorated with turning as were such other items as the bell standard. The side doors were decorated with painted designs, and the dial was covered with decorated, painted paper. The scapewheel consisted of a wooden disc with wires inserted in the face to take the place of teeth, and these impinged on the metal verge carrying the balance. The wood used to construct these clocks was beech, larch or mountain birch.

The trains were arranged one behind the other as on the iron clocks, and in some instances even quarter chimes were provided. When an alarm was fitted it was released by means of a peg inserted into one of a series of holes in the wheel carrying the hour hand. There were 24 of these holes, but the alarm could only be made to sound at a full hour or half hour–with no intermediate positions. The main dial only had an hour hand, but below it a small dial divided into four showed the quarter hours, and as the hand of this rotated once an hour, it fulfilled the same function as the minute hand on a modern clock.

The origin of most clocks of this type was Davos, although wooden clocks were made in many other parts of Switzerland. The typical Davos dial is a tall rectangle with a design cut at the top resembling a pair of horns flanking a *fleur de lis*.

Considering the rough construction of these clocks it is remarkable that they have lasted so long. They continued to be made in some places until the 19th century, but later examples were fitted with pendulums, mostly of the short "cow tail" variety swinging before the dial. Some of the clocks were even arranged to repeat when a cord was pulled. The achievements of Switzerland in the production of watches have overshadowed the clocks made in that country, but in these wooden clocks we have an admirable example of the results that can be achieved using simple materials.

Swiss wooden clock, 1669, regulated by altering the driving weight. The wheel balance is visible just below the top plate.

The Wooden Foliot Clock

TRADITIONALLY the earliest clocks in the Black Forest were based on a model brought in from Bohemia by a traveling glass salesman, probably in the second half of the 17th century. Owing to disturbed conditions in Europe at the time, there was no opportunity for the industry to establish itself until about 1720. Some of the earliest clocks have survived, however, and are interesting in that they were made by farming people with no technical background.

The movements were of wood. Very little metal was used, and this mainly in the form of wire. Only hard wood was suitable for making toothed wheels, and beech was the favorite although box is also found in some of these clocks. The teeth of the wheels were based on the isosceles triangle which was both strong and simple, but which limited the number of teeth that could be cut on the wheel. The pinions were lantern pinions made of two wooden discs connected by wires–thus forming a little cage which meshed with the wooden teeth. The pivots were wires stuck into the ends of the wooden arbors.

The escapement was of course the verge, and the scapewheel was formed from a disc with short pieces of wire stuck into its face. The verge itself was the largest metal component and carried a wooden foliot at the top which was provided with a number of notches to carry the small weights for adjusting the speed at which the clock ran. The driving weight of the clock was a stone held by a wire binding it, which was hooked on to the weight pulley.

Such clocks as these now seem of comparatively little use as timekeepers, but they served the needs of their time. They are important in leading the way to the large scale production of clocks in the Black Forest.

During the last few years, several firms in Europe have been making replicas of these clocks, and they certainly form a pleasing addition to the home. The clocks tick very leisurely –helping to suggest the quieter life of past centuries when the time needed to be correct only within a quarter of an hour or so. A further development of this idea is the clock kit, with which the purchaser constructs his own wooden wheeled clock. Modern methods of production are used involving plastic and plywood, while the teeth are machine cut and more accurate than they would have been on the old clocks. In addition, the teeth can be made smaller, with more modern materials.

Black Forest wooden clock, the earliest type known in this area. The date 1640 is merely a traditional marking.

Early Black Forest clock showing the simple movement. The foliot balance normally carries two small weights for regulation.

Foliot clock, modern reproduction, constructed from a kit.

The Onion

UNTIL the last quarter of the 17th century, the watch was incapable of accurate timekeeping. Workmanship had been improving and watches were being made smaller but until the invention of the balance spring they were of little value except as status symbols. The balance spring brought accuracy and also altered the sound of the watch; it caused a more rapid, rhythmical tick (previously watches had possessed a very irregular beat). The more rapid ticking meant that an extra wheel had to be added to the train to keep the watch running for its full period. Many of the early watches had only run 15 hours on a winding and now it was considered necessary to make them run a full day.

The tendency for the watch to become smaller reached an abrupt halt, but while in England the increase in size merely allowed for the extra wheel, in France watchmakers went to an extreme, making the watch as large as possible, and they produced a type that came

Opposite *This view of the watch gives an impression of its thickness. The watch is wound through the dial.*

Below *Onion watch by Thierry, Caen, late 17th century, showing the bridge used by European watchmakers to support the balance.*

to be known as the "onion", on account of its shape. It was claimed for the very thick watch that the parts were easier to make and the watch as a whole was easier to assemble. Larger parts were more accurate and longer arbors (axles) had a less detrimental effect on the meshing of the teeth and pinions as the pivot holes became worn. They could be made stronger, too, and therefore wore out less quickly. A larger watch would allow for a bigger barrel and give the mainspring more freedom so that it did not rub against the barrel bottom or the barrel cover and so on. All these claims led to two general conclusions: a large watch was easier for a watchmaker to deal with and would satisfy the customer by giving a better performance. The fashion of the time called for a thin watch, as the usual place to wear it was in the top pocket of the breeches. Enthusiasts for the thick watch claimed that it was just as easy to wear in the large pocket of the breeches–although this was bad from the point of view of the watch, as the movements of the wearer caused greater disturbance to it. A waistcoat was generally considered a much better place, as the watch was protected from movement and the wearer's body would act as a shock absorber.

Whether the public cared about scientific argument or not, the thick onion watch was fashionable in France from the late 17th century until about 1720. French makers liked to fit large balances to their watches and large scapewheels helped to maintain good timekeeping. The large balance meant that the watch had to be wound through the dial, a procedure which has always been more popular on the Continent than in England. Enamel dials were coming into favor at the time and on the onions such dials usually followed the prevailing fashion of a raised plaque for each figure.

Fashion eventually drove the thick watch away and later in the 18th century French watches became thinner than their English equivalent. Whatever its faults, it cannot be denied that the onion is easier to hold than a thinner, flatter watch and in a period when theatrical gestures were usual, this would facilitate the act of looking at the time.

Thirty hour long case clock with country-made case and brass dial bearing a single hand, early 18th century.

The Thirty-hour Long Case

THE lantern clock had its heyday in the first 75 years of the 17th century, but after the introduction of the pendulum, the invention of the anchor escapement and the establishment of the eight day long case clock, it began to lose popularity. London makers still produced it, but generally only as a small clock for travelers, fitted with alarm work only. While the country maker still kept the lantern in production, he usually fitted a larger dial imitating that of the long case clock, and he provided a wooden hood to protect the movement. The weight and pendulum still hung exposed below the clock.

By the early 18th century, the thirty hour movement was being fitted in a long case less elaborate than those fitted to the eight day movements produced in the larger towns. The dials of the thirty hour clocks were still square and possessed only an hour hand, but the separate chapter ring and spandrels were reminiscent of the eight day clocks. The cases were made by local carpenters, usually of oak or other local woods without much decoration.

The name of the "maker" is usually shown on the dial, but in many instances the place of business indicated is a very small village that could hardly support a clockmaker. It can therefore be inferred that these clock movements were produced in central workshops in towns and sold to village tradesmen with their names engraved on the dial by the workshop from whom the clock was ordered. The tradesman's normal business may have been that of blacksmith or ironmonger: in small communities a man had to undertake several trades in order to make a living.

By the middle of the 18th century, these smaller long case clocks were being produced with two hands and the cases were improving in quality. Many of the older one-handed clocks, sad to relate, have since been converted to two hands, but their original state can be detected by the fact that the dial only has quarter hour marks inside the figures and no minute marks outside them. Another unfortunate form of conversion has occurred when an old thirty hour dial has been fitted with an eight day movement, entailing the drilling of winding holes in the central matted surface of the dial. This work has often been clumsily performed and a pleasant clock has been ruined thereby.

There is no real disadvantage in possessing a thirty hour clock. Once the habit of winding

Thirty hour long case clock by Hedge of Colchester, Essex, walnut and pine, about 1785, a type made well into the 19th century.

it has been established, it becomes very little trouble and the performance is as good as that of an eight day clock. There is an advantage in having the endless rope system, for the clock is kept going while it is being wound, while an eight day clock stops for a few seconds, during which time the scapewheel may even turn in the reverse direction.

Even in times gone by, people regarded the thirty hour clock as an inferior piece. One often comes across these clocks with false winding holes painted on the dial, giving the impression that the clock is an eight day one.

The Birmingham factories producing clock movements during the late 18th century also turned their attention to the thirty hour clock. Some very good examples were produced, not only in the original birdcage form but also with plates similar to the eight day movements. Still another refinement copied from the eight day clocks was to fit a center arbor which would turn once per hour, instead of employing the more primitive method of driving the hands from an extension of the main arbor. This allowed the scapewheel to rotate in a clockwise direction and permitted the fitting of a seconds hand. Many of the earlier clocks did not have a pendulum beating 60 to the minute but a number like 58 or 64, and a seconds dial was not then possible.

In the early 19th century many thirty hour long case clocks with painted iron dials were made, and while they possess less elaborate cases than their eight day contemporaries, they are still very pleasing objects to have in the home.

The Act of Parliament

IN 1797 an Act was passed imposing a duty of five shillings per annum on any clock, ten shillings per annum on every gold watch and two and sixpence per annum on every watch cased in silver or base metal. The result was disastrous for the horological trade and many people were thrown out of work. Not surprisingly, the Act was repealed the following year.

The story has come down that as a result of the Act people disposed of their clocks and watches and relied on the tavern clock for their timekeeping; furthermore that the innkeepers anticipated this and had special clocks with large dials made for the benefit of their customers. This is fiction. Large mural timepieces were known as early as the beginning of the

18th century. H. Alan Lloyd in his *Collector's Dictionary of Clocks* illustrates one dated 1714 and a 1748 engraving, *The Art of Making Clocks and Watches,* illustrates this type of clock.

It seems likely that many taverns possessed clocks of this type at the time that the Act was passed, but an important fact to take into consideration when assessing the evidence concerning the Act of Parliament clock is its size. A clock of this type was intended for a very large room with a high ceiling. There are even instances of this type of clock being used in the interior of churches, as at Long Melford in Suffolk. Considering the size of the average 18th century tavern with its low ceilinged rooms, a clock of this type seems most unsuitable and leaves considerable doubt as to whether this clock was chosen by the innkeepers as the standard timekeeper, if indeed they made a choice at all. There are also too many of the clocks about for them all to have been made in the one year 1797–98.

The dials of these clocks were often painted black with gilt figures and the hands were usually of polished brass. The length of the minute hand made it necessary for it to carry a counterpoise, on account of the long overhanging weight needing to be balanced. The hand was therefore made as light as possible consistent with strength. The trunk of the case is often lacquered and finished with a design in Chinese style, a fashion that was current during the reign of George II. The trunk was usually long enough to take a seconds pendulum, but the fall of the weight was generally less than that of a normal eight day long case clock. The weight itself may have been shorter and wider than on one of the latter clocks in order to allow a few extra inches of fall. In order that the clock should go for eight days an extra wheel was necessary in the train to compensate for the shorter fall of the weight.

Later in the 18th century, slightly smaller versions of the mural timepiece were produced with dials covered by glass, but the spring driven clock became popular for large rooms and the weight driven mural timepiece was less fashionable in the early years of the 19th century. There was one main exception however: many railway stations in all parts of the country possessed clocks of this type which are now being replaced by electric clocks. There must have been a boom in manufacture when the railway system began to develop, for the clocks are found not only at important stations in the London suburbs and in provincial towns, but also in many smaller stations in the country.

Act of Parliament clock, inscribed "The gift of Sir Francis Forbes 1714", showing that the type was made before 1797.

Act of Parliament clock by Thomas Moore of Ipswich, Suffolk, mid-18th century. The drinking scene is on at least two other Moore clocks.

English dial, late 18th century. The "salt box" case is of oak with a mahogany dial surround bearing a rope type decoration.

English trunk dial in mahogany case by James Bannister, Leicester Square, London, 1850. The dial itself is flat but the glass is convex.

The English Dial

THE mural timepieces known today as Act of Parliament clocks were suitable for large rooms such as were found in inns and public buildings, but towards the end of the 18th century it became fashionable to produce a slightly smaller clock with a light colored dial in place of the previous black ones. At first the dials were made of brass and silvered over and engraved. Later the painted iron dial was used as on contemporary long case clocks. Dials were by this time usually covered with a glass.

As a further stage of development, these clocks were made smaller still and became spring driven. This eliminated the long trunk to contain the weight, and the whole case was hidden behind the dial–hence the type became known as "office dial" or often just "dial". The movements were based on those of the bracket clocks of the time, had a verge escapement and of course the inevitable fusee contained between tapering plates. The tapered plates were retained until well into the 19th century and until after the verge escapement had been replaced by the anchor escapement.

The iron dials were often made convex and the glass, of course, had to be made convex as well. The bezels on the earliest clocks were of wood, but later brass bezels were used, very solid at first but becoming lighter as the 19th century wore on. While the earliest clocks had locks to hold the bezel in position, the later ones had only a catch.

The usual wood for the cases was mahogany, but oak cases are seen on clocks made in the 20th century. At first the surround to the dial was screwed to the case, but later it was found more convenient to have it fastened with wooden pegs so that the dial and movement could be easily removed. A small door in the side of the case allowed the movement to be inspected and the bottom of the case followed the outline of the dial's wooden surround, requiring a curved door to give access to the pendulum.

Although the dial clock began its existence as a scaled down version of a larger clock, once the design was established some models were made once more with a trunk, although this was very short compared to the early weight driven mural clocks. The trunk allowed for the advantage of a longer pendulum and also gave scope for more decoration, such as brass inlay.

The standard size for an English dial is 12 inches in diameter, but larger sizes have been made for special use and smaller ones are also known, although much rarer. An 8 inch dial clock with a trunk case is a very attractive object, but a genuine old one is hard to find.

English trunk dial by Ainsworth Thwaites, late 18th century. Characteristic are the heavy bezel and the "ear pieces" below it.

Repeating watch with cylinder escapement in gold case by Lacroix, Turin, 1833, and repeating watch by Breguet, about 1800.

The Repeater Watch

THE application of the balance spring to watches in the third quarter of the 17th century transformed them from expensive toys into timekeepers accurate to within a minute or two a day. Soon afterwards the invention of the rack and snail striking mechanism for clocks ensured that a clock could be made to strike at will and would always sound the correct number of blows in accordance with the time indicated by the hour hand.

Technology was rapidly developing at the time and two London makers, Tompion and Quare, decided to combine these two features to produce watches that would sound the time at any hour of the day or night–which would be extremely useful when artificial light was difficult to get at a moment's notice. More than merely striking the hour, they made their watches strike the quarters. The two watches were shown to the King, James II, in 1687 and he preferred Quare's because it was only necessary to press one knob to sound both hours and quarters, while on Tompion's a separate knob had to be pressed for each. The watches were, of course, fitted with the verge escapement, which made them thick, but the entire mechanism was incorporated without greatly increasing the size beyond that of a normal contemporary watch. Watches had already been made that struck the hours in regular progression and these had their bell surrounding the movement. The repeaters were made on the same plan.

Once the principle of repeating watches became known, London makers developed the idea and it spread from England to Europe. The next development was to ensure that the correct time was struck, because if the knob were not pushed in far enough too few blows would be sounded. The mechanism was therefore arranged not to sound at all unless the knob had been pushed in sufficiently.

Quarter repeating watch with lever escapement, additional chronograph and stem wind, a product of the late 19th century.

Dial view of the watch showing the center seconds hand for the chronograph in addition to the normal one.

By about 1730, repeaters were being made to sound half quarters ($7\frac{1}{2}$ minutes) and five minute repeaters were also being made. A century later, watches were even produced to repeat the time to the nearest minute and this feature could also be found on very small watches made towards the end of the 19th century. A further refinement was the pulse piece, whereby the blows of the hammers were transferred to a knob in the side of the case and the owner of the watch could feel the blows instead of listening to the strokes on the bell.

Later the fashion for smaller watches led to the omission of the bell and the hammers were made to strike on metal blocks producing a so-called "dumb repeater". The dumb repeater was not very popular, but the problem of size was solved by fitting repeating watches with wire gongs that surrounded the movement and made a more musical sound without taking up the space occupied by a bell. When during the late 19th century, keyless work was introduced, the old method of pushing the pendant to make the watch strike was abandoned in favor of a slide on the side of the case.

The repeater was expensive to make and to repair. It owed its demise to the invention of electric light and the use of luminous dials. It has, however, been immortalized in literature by Charles Dickens who describes Scrooge using one and it has also formed the basis of a scene in the opera *Die Fledermaus*.

The Breguet Souscription

THE great French watchmaker, A. L. Breguet, specialized in producing high class watches to order. Large numbers of these exist today showing great differences in design and additional features over and above simple timekeeping. All exhibit the exquisite workmanship associated with his name.

Below *Pair of* Souscription *watches by Breguet, nos 445 and 1350. The extreme simplicity of the movements can be seen here.*

Opposite Souscription *watch dial. The winding square is at the center and the hand has a very fine tip for reading the time accurately.*

While producing these costly pieces, he aimed to make a cheaper watch by production in batches. He designed a very simple movement with fewer parts than usual and provided it with only one hand, although two handed watches were the standard at that period. The watch was slightly larger than usual ($2\frac{5}{16}$ inches diameter) and divisions were made between the hour figures, allowing the time to be read to the nearest five minutes, so that with practice an estimate to the nearest minute could be made. The plain dial and small Arabic figures aided legibility. The winding square was placed in the centre of the dial and was surrounded by a tube carrying the hand–which saved making an extra hole in the dial, for most European watches of the period were arranged to wind through the dial.

While these watches were cheaper than the usual Breguet watch, the word "cheap" is only relative. The price was from 600 francs and in order to finance production and allow several to be made at one time, prospective customers were expected to pay one quarter of the price in advance. The basic cases were of silver with simple engine-turned decoration and gold bezels, although variations on this theme do occur. The exterior of the watch was very plain. The single hand had a long extension for reading divisions of an hour and was set simply by pushing it around with some small object. There would be a temptation to use the winding key for this purpose, with the risk of scratching the enamel dial, and if the owner were at all clumsy the hand was easily broken.

It was Breguet's practice to sign his watches

with a secret signature, that is, a signature so small that it was scarcely visible without a lens. The *Souscription* watches have inscribed on them, in addition to the name Breguet, the word *Souscription* and the serial number. The cylinder escapement was used to save space, and was made in Breguet's special form, in which the cylinder came at the end of the staff with the scapewheel placed immediately behind the dial. An added refinement was a "parachute" or flexible bearing for the balance staff, which protected it from shocks if the watch should be accidentally dropped. Similar devices are used on modern watches, but the present day use of this device is only a few decades old. The balance was made of gold.

By simplifying the mechanism, Breguet believed that the watches he made could be easily repaired in any country. There was no fusee included, but the mainspring was made to run for 36 hours whereby a more even driving force was given to the mechanism. About 150 *Souscriptions* were made by the Breguet firm from the late 18th to the first quarter of the 19th century. Other makers tried to get a share of the market and produced similar watches which, although resembling the design, lacked the refinement of the original.

Comtoise in use as a wall clock. The pressed brass surround to the dial is a well known feature.

Highly decorated pressed brass pendulum typical of those fitted to Comtoise clocks. These pendulums sometimes have a painted decoration.

The Comtoise

THE roman clock was first made in the early 18th century in the village of Morez in the Morbier district of Franche Comté. All these names are applied to the clock, which was made with little variation until 1914. The movement consisted of vertical iron bars with brass bushes for the pivots, and the top and bottom were made of sheet iron which was also used for the sides and back of the case. The dial usually consisted of a circular enamel portion containing the figures surrounded by pressed brass and with a pressed brass ornament above. Sometimes fanciful shapes were used.

The clocks penetrated to all parts of France, and many people went on to have a long case made for them by a local cabinet maker. Consequently, the case styles reflect local characteristics. Sometimes the clocks were used simply as wall clocks.

"Vineyard" clock. Such clocks are often fitted with spring driven Comtoise movements.

The movements were well made and the steel pivots very hard, ensuring long life. A peculiarity of these clocks was the striking mechanism which repeated the hour two minutes later, but this feature was not introduced when these clocks began to be made, for it had already been known for centuries. The theory was that if one missed the first striking, one was on the alert to count the strokes the second time. The earlier clocks were fitted with the verge escapement and the scape-wheel upside down, but with long levers to keep the amplitude of the pendulum small. The earlier clocks had the pendulum at the rear, but more modern examples had theirs at the front immediately behind the dial. The last clocks made also had the anchor escapement instead of the verge. Sometimes alarm work was fitted, in addition to the striking.

It is possible to date these clocks very roughly by the features of the dial: for example, those made about the time of the Revolution have the revolutionary cap and the cockerel, while a date before 1789 or after 1815 can be inferred from a clock of *fleur de lis* design.

The earliest dials were made entirely of metal and often bore a garter with a motto below the chapter ring that had to be pushed aside to reveal the winding holes. The earliest pendulums were made in sections jointed together, some say for ease of transport, others, to give a little play and to help the action of the verge escapement. The bobs were shaped rather like the handle of a poker.

Perhaps the most remarkable feature of later clocks is the highly decorated pendulum of pressed brass, as in the clock shown opposite. Sometimes these pendulums in addition are painted in bright colors and even have automata in the bob.

These clocks have always been popular with European collectors, but have only become well known in Britain in the last 20 years. Their popularity is increasing so much that a firm in France is now making reproductions closely following the original design, although the anchor escapement is used.

Spring driven versions of the movement were also made, and these are often found in the "vineyard" clocks that are beginning to be imported into Britain. The same solidity of construction is found in these movements but the spring driven variety almost without exception have the anchor escapement.

The Pocket Chronometer

THE chronometer was developed for the purpose of finding longitude at sea. A navigator can calculate his distance east or west of Greenwich if he compares the local time obtained from an observation of the sun's altitude–which when at its maximum is the local noon–with Greenwich time. Greenwich time is supplied by the chronometer, a large watch mounted in gimbals to keep it steady as the ship rolls, and specially constructed to vary as little as possible during long voyages. Differences in temperature cause the balance to expand or contract and alter the elasticity of the balance spring–all of which could effect the rate of the timekeeper. The chronometer is specially constructed so that these difficulties are overcome as nearly as possible. To further assist accuracy, the chronometer has a fusee to give a steady driving force throughout the

Pocket chronometer by Frodsham, mid-19th century. The appearance scarcely differs from that of a lever watch of the same period.

Pocket chronometer movement by John Arnold and Son, late 18th century. The balance is compensated for temperature.

period of going and a special escapement to enable the balance to vibrate with as little interference from the train as possible.

The first successful marine chronometer was made by John Harrison, a carpenter from Barrow on Humber, but this machine was too large and complicated to be workable and he constructed three more before he produced one that was really practical. Even this, which was in the form of a watch 7 inches in diameter, was a complicated piece of mechanism and could not be produced cheaply.

The men who developed the marine chronometer in its present form were John Arnold and Thomas Earnshaw. Once they had produced instruments suitable for use at sea they turned their attention to a similar timekeeper for pocket use. The principles of the pocket chronometer are similar to those of the marine type, but it is cased in a metal case like that of a watch. It suffers certain disadvantages as it does not lie on its back supported by gimbals to keep it steady, and it is held in an upright position in the wearer's pocket, thus being subjected to disturbance every time he moves. The horizontal position of a marine chronometer means that the weight of the balance is carried by the end of one of its pivots, whereas the weight of the balance of a pocket chronometer is borne by the sides of both pivots, thereby increasing friction.

The balance of the chronometer is only given impulse at every other vibration and it is possible that a sudden movement on the part of the wearer of a pocket chronometer may stop it. In practice, pocket chronometers have been used successfully in place of watches and have given a good account of themselves, but they are difficult to make and repair and are consequently expensive. It is inconvenient for the owner of a pocket chronometer to take it to the watchmaker every time it needs adjusting and it is doubtful whether the extra accuracy obtained from the instrument is worth the extra trouble. A marine chronometer should never have its hands altered, nor should it be adjusted for regulation except by an experienced craftsman. A good lever watch will give results sufficiently accurate for normal life. Modern quartz watches with a much greater degree of accuracy render a pocket chronometer unnecessary and even threaten the supremacy of the marine instrument.

The Black Forest Wall Clock

CLOCKMAKING in the Black Forest traditionally began in the 17th century and by the end of the 18th production was high. Clocks were being sold all over Europe and even in America. Development of the clocks had been slow, but they eventually sold well because they were cheap, reasonably accurate and ran for years without attention.

They were usually sold by traveling salesmen, who were often away from their homes for years at a time on the road, but in the 19th century the practice developed of establishing groups of Black Foresters in far away places who kept in touch with their home villages and organized the sale of the clocks from their new base.

What might be called the standard Black Forest clock produced in the early 19th century had a frame made usually of beechwood. The brass wheels were mounted on wooden arbors with wire pivots that worked in brass bushes mounted in the wooden frames. The wooden arbors were often covered with bronze or silver paint so that they gave an impression of metal. The pendulums were up to three feet long and the dials and side doors of the clocks were made of fir or pine. The doors were usually covered with paper, but, being made of coniferous wood, became brittle and have often been lost. Dials are often found in a split condition, usually with some unskillful repair work.

Black Forest clock peddlers—a scene showing the workshop, a village dance and a traveling salesman.

Painted arch dial, more typical of Black Forest clockmaking than of the type made for the British market.

Typical Black Forest clock made for the English market. Some attempt was made to imitate the English dial.

The dial of the standard type is usually about 10 inches square with an arch above. The wood is painted white, decorated with colored flower designs and afterwards varnished. Below the clock are iron weights hanging on brass chains. For the British market, the dials were usually made circular with a mahogany rim and covered by a domed glass in a brass bezel.

The clocks were made originally with the striking train near the wall and the going train in front of it, but in course of time the frame tended to sag and the pivots would bind in their holes, so the practice was later adopted of placing the trains side by side. Smaller versions of the clocks were made, some with porcelain dials with tiny movements or wooden frames.

Once the American OG (short for "ogive") clocks began to penetrate world markets, the popularity of Black Forest clocks gradually receded, and Germany was forced to make clocks in factories similar to the American ones. Nevertheless, the traditional Black Forest type remained alive in the so-called "postman's alarm". These clocks were like the striking clocks but had alarm work placed at the rear, where the striking train had formerly been located. Tradition died hard, for the fore and aft position was retained, but the wooden arbors were replaced by metal ones. These clocks were still being sold as late as 1914 and successfully competed with the spring driven alarms produced in Germany and America.

Black Forest clocks are now becoming collectors' items. They suffer from the disadvantage of being susceptible to woodworm, but even those in the most pathetic condition are usually capable of restoration. The type has one great advantage for collectors: the clocks were so cheap when first made that no one has bothered to produce any fakes.

Typical Black Forest movement, with the striking train placed behind the going train and the bell above.

Below *Farmer's verge movements, typical early 19th century. The balance cock of the movement on the right tends towards plainness.*

Bottom *Farmer's verge in a silver case showing "bullseye" glass, and the square for setting the hands by the key.*

The Farmer's Verge

THE verge escapement, employed in the earliest watches, continued to be used almost throughout the whole period during which watches were wound with a key. Small verge movements were made during the 18th century, but most of those of the 19th tended to be large, containing much less decoration than the verge watches of earlier times. The balance cock was of the plainer type usually associated with the lever watch and the balance itself now had a much broader rim than before.

The origin of the term "farmer's verge" is obscure. There is no evidence to suggest that the type was worn particularly by farmers, but it is probable that watches of this type would enjoy a better sale in country towns still dependent on agriculture than in centers where the Industrial Revolution was well advanced.

A typical verge watch of about 1830–40 might have been as much as $2\frac{3}{4}$ inches in diameter and nearly $1\frac{3}{4}$ inches thick including the dome of the glass. The latter would probably be of the bulls-eye type, so that the watch could be laid dial downwards when it was not being carried in the pocket. The case would be a silver pair case, and the movement contained in the inner case would lie in the outer case on a bed consisting of one or more watch papers – circular advertisements of the watchmaker who supplied the watch or who had later cleaned or repaired it. Sometimes, a girl would work little circular embroidered designs for her fiancé to put in the back of his watch instead of papers. The dials of these watches often had colored pictures on them and one has even been noted with a picture of an early railway train.

At the time when these watches were made, other watches were also large and it would only be after the smaller fullplate lever watch began to be popular that the farmer's verge would appear old fashioned and rustic. They were still luxurious commodities at the end of the 18th century, but at the beginning of the 19th, several new escapements were coming into use which gave much more accurate results, so the verge declined in popularity. The fashion for thinner watches and the use of keyless work, which created difficulties when applied to a fusee, also helped to kill the verge watch, for a verge escapement has to be used in conjunction with a fusee if any sort of timekeeping is to be expected from it.

While the watches would have been considered obsolete at the time when they were made, they would have been capable of telling the time sufficiently accurately in a community not yet fully dependent on railways, and they were strong enough to stand up to hard knocks. Their movements were reasonably large, making for ease in stripping and assembling during repairs, and they also had the advantage of being a type that the trade was used to and had had experience of for the previous 150 years. A country repairer of those days might well have spoilt a lever watch because of his unfamiliarity with the niceties of the escapement.

The Pendule de Paris

THE spring driven pendulum clock began to be made in France shortly after the pioneer examples had been produced by Salomon Coster in The Hague in 1657. The earliest French clocks were based very closely on the Dutch models, but as the century progressed, the furniture designs of Boulle and other workers led to new cases being designed which were more elaborate than the original ones, whose plainness had caused the French to call them *religieuse* (nun).

In 18th century France the clock was firstly an article of furniture and only secondly an object of utility. The French striking clock movement, therefore, did not vary greatly throughout the century, although the exterior cases show remarkable changes. By the time of Louis XVI, movements and cases were being made more delicately and the movement tended to have circular plates, whereas previously the plates had been rectangular. By the early 19th century the circular movement had established itself, and in 1810 an important event in its history took place when Frédéric Japy (1749–1812) established a factory for clock movements at Badevel. The firm received a gold medal in 1849 and exhibited at the great exhibition of 1851 held in Hyde Park, London. Many clocks are now seen in Britain with the Japy stamp on their back plates.

Typical French movement of the first half of the 19th century with silk suspension for the pendulum adjusted by the milled knob.

Japy was not the only company in the clock industry. The French striking movement was turned out in bulk by many factories, but most of the examples noted exhibit the same high finish and capacity for working for years without attention. Fashion in cases changed, but the old type of movement continued to be produced. In the 1850s it was fashionable to have clocks in gilded cases covered by a glass shade. The later Victorian period favoured black marble cases for these clocks; later, green onyx cases decorated with bronze mounts were in vogue. The fashion began in the 1860s to reproduce designs of the previous century with the typical circular movement.

What we may call the basic design of the movement had a countwheel for controlling the striking, and a loop of silk that was adjustable in length for supporting the pendulum. Later improvements included Achille Brocot's steel suspension for the pendulum, and rack striking, whereby the clock always struck in

Below *Group of* pendules de Paris *showing the widely differing types of case into which the typical French movement could be fitted.*

Bottom *French movement in a case suggested by the Gothic revival, about 1830. Later, less tall cases were to allow for shorter pendulums.*

accordance with the time shown by the hour hand. The Brocot suspension could be regulated from above the dial, and it was therefore unnecessary to shift the clock for this purpose – an important consideration when the weight of the heavy black marble case needed two people to shift it.

Other improvements to the movement were a visible escapement with pallets made of agate, invented by the father of Achille Brocot, and calendar work operated by the striking train.

The French movement became so popular that imitators naturally stepped in. Ansonia and Waterbury in America, among others, produced their version of the clocks in black marble cases (sometimes iron or wood painted to look like marble), and even had their own version of the visible escapements.

While the gilded cases appeal to collectors, the black marble versions do not, and many of them have disappeared in recent years. This is a great pity, for the French movement is one of the best produced on a large scale. In the early years of the century it was possible to buy the complete clock in London for prices from 1 guinea upwards for a timepiece, and between £2 and £3 for the same clock fitted with striking mechanism. As the cases increased in size, so did the price. It is possible that one day the black marble clocks may become popular once more.

This type of case with pillars and entablature was very popular on French clocks of the First Empire period.

Dial of a Yorkshire long case clock by Scholefield of Dewsbury. The painting probably represents the infant Samuel.

Full-length view of the Scholefield clock. Note the elaborately turned pillars flanking the hood and the small door in the trunk.

The Yorkshire Long Case

BY the second half of the 18th century the long case clock was going out of fashion in London, but was still popular in the provinces. In Wales and the North of England the type was still in demand, and movements and dials were being manufactured in Birmingham and sent to various parts of the country with the local clockmaker's name inscribed on the dial to imply that he was the maker of the clock. A few makers made their own movements and fitted Birmingham dials to them. This does not mean that the clocks were inferior to the hand made product, for they possessed solid brass movements of the conventional type that were capable of standing up to long use and giving a good performance.

With the Industrial Revolution and the rise of rich mill-owners in Lancashire in the late 18th century, it became fashionable for them to provide their homes with large long case clocks, many of which had special features like the phases of the moon or center seconds hands. The dials of these clocks were often covered with engraving and, as they possessed extra hands to indicate the additional information, they were not easy to read at a distance. The pleasing outline of the London long case clock of the late 17th century completely disappeared: the cases were now decorated with carving, sometimes in the form of imitation brickwork and other devices.

On the other side of the Pennines, in the woollen districts of Yorkshire, much the same thing was going on, but here the fashion was extremely wide cases with a very short trunk so that the door giving access to the pendulum and weights was nearly square. The cases were generally of rich mahogany, and included paneling, turned pillars and other forms of ornament. The dials were now of painted iron, while in Lancashire clocks with brass dials decorated by engraving were favored. The dial paintings of the Yorkshire clocks are often most interesting, in that certain designs appear to contain some special meaning of local or allegorical significance. Their hoods were made very large and the top bore a scroll design.

The hands of a Yorkshire clock were usually pressed out of sheet brass and formed a matching pair. The seconds and day of the month hands usually followed a similar design to the hour and minute hands. The very large base was supported by feet, which in later examples were turned.

The movements call for little comment, being typical of movements found in other parts of the country. It is the case that makes the Yorkshire clock. Judging by the names found on the dials, they were made up to about 1840 and, strange as it may seem, while most of the dials bear the names of Yorkshire towns such as Bradford, Dewsbury and Otley, clocks in the same style of case can be found bearing the names of towns many miles away. The most distant place name observed so far was Limerick, Ireland, and for good measure this clock, although a typical Yorkshire clock in appearance, had a thirty hour movement and dummy winding holes. The dummy winding holes have also been seen on a large clock from Dewsbury, so it seems that there were poor relations to be found in Yorkshire itself.

The Yorkshire clock has never been popular with serious collectors. However, it has great practical value: its large, broad case gives it the advantage of a good firm support for the movement that makes for steady timekeeping.

The Banjo Clock

TWENTY-FIVE years after the signing of the Declaration of Independence, clockmaking in the USA, although based on British traditions, had begun to break away into new styles. This was due to a shortage of raw materials and also of skilled labor, while at the same time immigrant clockmakers from other parts of Europe were ready to influence fashion by perpetuating ideas from their native countries.

A clock smaller than the English long case type was beginning to be made during the last years of the 18th century. It gave the impression of a spring driven clock standing on a small cabinet, but in reality the two parts of the case were one and their combined height provided a space through which the weight could fall. The gearing was arranged so that the clock could run with a much shorter weight fall than a standard long case movement.

In 1802 Simon Willard of Roxbury, Massachusetts, patented a new type of clock that was the ultimate development of this idea. The term "banjo" is derived from the shape of the clock and is a more modern coinage, for the clock type was initially known as the "patent timepiece". The top of the case which housed the movement was circular and slightly larger than the dial, and this was supported by a

Banjo clock by Aaron Willard, Jr, 1841. Later examples of this type of clock tend to be more ornate than the original version.

tapering centre section of the case which stood on a rectangular base. The centre section contained a tapered weight which had a very limited fall and therefore had to be heavy in proportion to the size of the clock. The weight was compounded, that is supported by a pulley so that the amount of line running through the clock was equal to twice the distance through which the weight fell. An extra inch or so of space was obtained by locating the barrel opposite figure 2 on the dial, thus allowing the pulley to rise higher than if the normal position at figure 6 had been chosen. The pendulum hung between the movement and the dial in order to save space and was formed with a stirrup to clear the tubes carrying the hands. The wheels and frame were similar to those used in English clocks, but scaled down. The pendulum bob swung in the rectangular section at the bottom of the case, separated from the weight by a thin partition. This was necessary because as the weight descended, the time would come when it would form a pendulum having the same period of swing as the pendulum of the clock. It would then start swinging on its own and probably cause the clock to stop.

The clock was decorated in gilt with polished brass ornaments at the side, and originally had "S. Willard Patent" painted on the glass of the rectangular portion. The glass of the tapered section in the center was also painted, but with a conventional design.

The clock as visualized by Willard had a plain acorn finial, but other makers produced their own versions with other finials such as bronze eagles, and later painted tablets were used for the rectangular portion below. These clocks are very desirable collectors' items in the USA and therefore any specimen should be examined with great care for possible cannibalization or modern additions.

Later developments were a lyre-shaped center portion or a circular base; in the latter case the clock was then known as a *girandole*. However, the original form, in the opinion of many collectors, cannot be bettered.

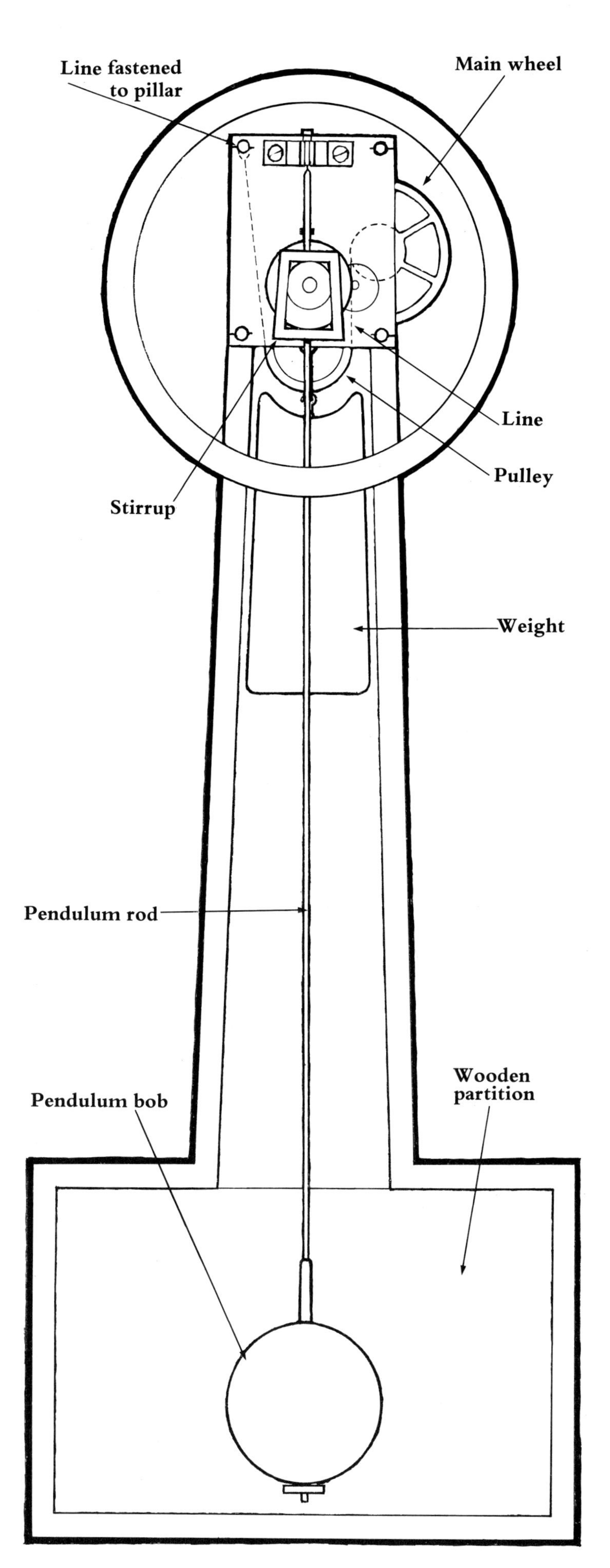

General arrangement of banjo clock, showing position of weight and pendulum.

English fullplate lever watch by John Ballard, Lamberhurst, Kent, 1847. The seconds dial flush with the main dial signifies early work.

Very early lever watches have their seconds dial flush with the main dial, but later the seconds dial was sunk below the surface to give more clearance between the hour and seconds hands: these had been set closer together as the watch became thinner. Some balances were of plain steel, others were of gold, while the first quality had brass and steel compensation balances with timing screws in the rim.

The watches took a long time to strip and assemble, a process in which the fusee was always a drawback. Some of the watches made were originally intended to have fusees and yet were fitted with extra empty going barrels that merely transferred the power of the spring to the train, without doing anything more useful than occupying the space where the fusee should have been. This was an economy measure caused by competition from America and Switzerland–competition that was seriously threatening the continuation of the English watch industry. Eventually the fullplate watch was discontinued in favor of the three-quarter plate, but by then it was too late.

The English Fullplate Lever

THE lever escapement was invented by Thomas Mudge, a London watchmaker, in the 18th century, but he showed very little interest in his device and it was left to other makers to undertake the necessary work of developing it. The lever escapement differed from other escapements in use at the time in that the balance was free for most of its swing and only made contact with the rest of the mechanism for a short period at each vibration. It was capable of standing up to everyday life better than those escapements that received impulse on every other beat, such as the duplex, and once the various difficulties had been overcome, the lever watch became the most popular English watch of the 19th century.

The earliest were as large as contemporary verges of the late 18th century, but the makers strove during the first half of the 19th century to make the watch smaller, even though the fusee was retained. In addition to retaining the fusee, which now needed an even finer chain than before, a device was incorporated in the fusee to provide a little power in order to keep the watch going while it was being wound. The old pair case of the verge watch was replaced by a new one with an inner dome containing the winding hole, although the hands were still set and the movement still swung out of the case from the front.

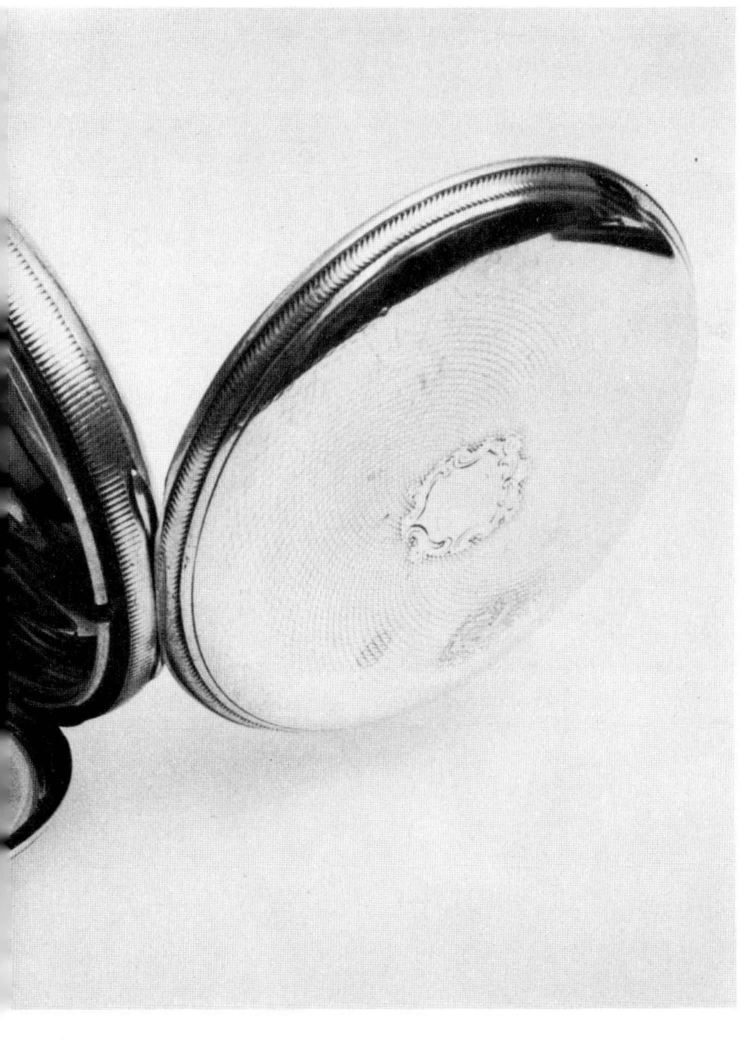

By the early years of the present century the English watch industry was virtually dead. Although the fullplate lever was an excellent product it cost too much to make when compared with European and American watches made without fusees and put together by semi-skilled labor in factories producing large quantities of standard parts. True, the English watch contained much mass produced material but there was still too much hand finishing involved to allow the English product to compete with imported watches. Moreover, the imported watches favored keyless work, could be made thinner and were made with improved scapewheels that were stronger and held the oil better than the English watch.

There are still many fullplate levers in existence, but it is difficult to find the necessary skilled labor to repair them. They are scarcely collectors' items yet, but it may not be long before they attract interest.

Above *English fullplate lever movements. The top movement is typical, while the two by J. G. Graves have dummy fusees and the one by Dent is a de luxe model.*

Left *English lever watch case construction. The glass opens to adjust the hands and the back door reveals the winding square when opened.*

Early American OG, about 1843. The dial is of wood with colored designs and the hole in the center allows the escapement to be seen.

American double OG by the Waterbury Clock Co., about 1880. By this time, the dial was being made of sheet zinc.

The paper was typical of early American clocks. The wire gong for the striking and the hook-on pendulum bob can be seen here.

The American OG

MASS production of clocks in America began in the early 19th century with a wall clock based generally on the English long case movement. The American movement was almost entirely of wood and such clocks were made in very large numbers in factories powered by water wheels. About the year 1815, Eli Terry, the pioneer of this production system, devised a new clock with a wooden movement in which the weights were enclosed in the case; the clock was therefore able to stand on a shelf instead of having to hang on the wall. Other manufacturers imitated the design: Chauncey Jerome, had the idea of creating a cheap shelf clock with a brass movement. His brother, Noble Jerome, designed and patented such a movement in 1839.

The case for this movement was rectangular – about 26 by 15 inches – and the door in the front, which was provided with clear glass in front of the dial and a painted tablet below, was set back and surrounded by a moulding based on the ogive. This moulding gave its name to the style, and the clocks were known as "OGs". Sometimes the outer edge of the case was decorated with a convex moulding about $\frac{3}{4}$ inch wide and the door mouldings were concave, repeating the ogive motif. This pattern was known as a "double ogee" or "OOG".

A wooden wheel naturally had fairly large teeth, and consequently the number of its teeth were limited. As a result, five wheels were needed in the wooden movements for each train, but the Jerome brass clock took full advantage of the material and used only three arbors on each side, combined with large wheels having a number of teeth much greater than could be used on a wooden wheel of the same size. The movement was a masterpiece of simplicity, lending itself well to mass production. Many American firms produced it, forming the basis of the export trade in American clocks that spread over the entire world. When the Black Forest industry turned over to making copies of American clocks this model was also included. Production in the USA lasted from about 1840 to 1914 but in Germany for a shorter period as the German factories branched out into designs of their own.

The inside of the case usually contained a paper giving the maker's name and instructions for keeping the clock in order.

The American Fullplate Watch

BEFORE the middle of the 19th century, there were not many watches made in the USA. Most watches used had to be imported and Liverpool in particular sent over a large number.

In the early 1840s, factories in Connecticut were turning out large numbers of clocks by machinery and even exporting them. Soon it occured to a Boston watchmaker, Aaron L. Dennison, that watches could be made in the same way. He began experimenting in 1848 in company with Edward Howard, a clock-maker, and Samuel Curtis, financier, starting a small factory at Roxbury, Massachusetts in 1850. Four years later the factory was moved to Waltham, but in 1857 the company failed and was purchased for Appleton Tracy and Co., the name being changed later to The American Watch Company of Waltham, Massachusetts.

Many other people started watch factories in the USA and a great number of these companies failed, but some are still functioning today.

The early factories concentrated on watches based on the English fullplate type with a lever escapement, but with the important difference that the fusee was omitted. Various grades of movements were produced, some containing more jewels than others. The better models were provided with compensation balances, while the cheaper ones possessed plain balances only. The balance cocks contained a certain amount of decoration, but otherwise no money was wasted on adornment. The maker's name would be engraved in script or block letters. It was usual to name different models after the various directors of the company. Spare parts were always available and catalogues of these were issued.

The usual practice was for the factories to make the movements only which would then be sold wholesale to other companies who would provide cases. The case would need a hole in the inner cover at the rear to allow for winding, but hand setting was done by means of a square on the minute hand as on an English full plate watch. To facilitate this, the bezel had to be capable of opening easily.

It became fashionable to fit these watches with a safety pinion on the center arbor, and to state this on the movement. This safety pinion worked on the principle of a center pinion screwed on to its arbor; in the event of the mainspring breaking, the pinion would become unscrewed and no damage would be done. (The kick when the spring broke was enough to break several leaves off the center pinion, necessitating an expensive repair.)

The fullplate movement lasted virtually until the end of the 19th century, but the desire for smaller watches and the use of keyless work caused the factories to design three-quarter plate models and to produce smaller watches. The basic design of the full plate must have been good, for even today there are still some in regular use as they approach their centenary.

American watches can be approximately dated by their serial number. While American libraries should contain reference books dealing with this matter, the only book available in Britain is by Lt Col George Townsend: *Almost Everything You wanted to Know about American Watches and didn't know Who to Ask.*

Opposite *Two grades of Waltham movement: Farringdon H with compensated balance, about 1890, and Farringdon D, about 1885.*

Below *American fullplate movement shown in its case and from the dial side, Waltham, Massachusetts, about 1885.*

Early Vienna regulator. The dial is plain and the hands slender while the decoration of the case is restrained.

Typical Vienna regulator as produced by German factories in the late 19th century. The dial has a center brass rim.

The Vienna Regulator

THE Vienna regulator developed in the early years of the 19th century, and in its earliest form was like an English long case clock with a distinctive hood, trunk and base. However, its glazed case distinguished it from a long case clock, and the movements were always more delicate. While some models were made to stand on the floor, this type has always been essentially a wall clock.

The clocks were produced by masters in Vienna during the Biedermeier period and were made by hand, except for the cutting of the wheels. The high finish of the movements meant that only a light weight was necessary to drive them and the comparatively heavy pendulum with the dead beat escapement helped to make a very accurate clock. The pendulum rods were usually made of a wood well dried and treated with various preparations to make it waterproof and virtually immune to changes in temperature.

The clocks appeared as timepieces and with hour strike–sometimes even quarter chimes. Some were fitted with a cord that could be pulled at night to make the clock repeat. A genuine Austrian example usually had its movement mounted on a wooden seat board and bore the name of its maker. In Austria the clocks were used as railroad timekeepers and also for government offices and institutions because of their accuracy. Even the Emperor had one in the study of his country house at Bad Ischl. The composer Franz Léhar had a miniature one about 2 feet tall in his home.

When Germany began to develop its clock industry on a factory basis in the mid 19th century, the Vienna regulator became a favorite item for production. The quality of the German factory clocks was generally good, but they did not have as fine a finish as the genuine products of most of the larger Austrian companies. In Germany, the products of Lenzkirch A.G. are now very highly regarded. A feature of the German clocks was the seconds hand, which moved round a dial calibrated to 60 but actually taking less than a minute to do so.

While the majority of the German clocks were made in the Black Forest area, most of those seen in Britain were made by Gustav Becker of Freiburg. These clocks bear his trademark–an anchor and the letters G.B. This Freiburg is not Freiburg im Breisgau, Baden, but Freiburg, Lower Saxony, and other factories located in this town were eventually amalgamated with the Becker concern.

Vienna regulator by Gustav Becker, Freiburg, Lower Saxony. The Arabic figures are unusual on this type of clock.

A popular product always has its cheaper imitations, and the Vienna regulator was no exception. The German factories produced some clocks which were slightly smaller and are usually known simply as "regulators". They had the same style of case, but the pendulum was usually a gridiron of brass and steel rods which were only for show and not arranged to give temperature compensation. The clocks were spring driven with anchor (recoil) escapements. There were two qualities, one with solid plates and the springs contained in barrels, the other with the stamped out plates and open springs of the American type. About the turn of the century, smaller examples were also produced.

Even American factories brought out their own versions of the Vienna regulator, both weight and spring driven, nearly all of the large size.

The Waterbury Company could easily adapt timepieces to alarms by fitting the separate alarm mechanism shown.

The Cottage Clock

THE early American factory-produced clocks were all weight driven. It was not until the mid-1840s that the problem of producing steel springs commercially and cheaply had been overcome, but once the spring was available as motive power designs multiplied rapidly and all factories strove to bring new models on the market to outsell their competitors. The earliest spring clocks were mostly strikers, but it became fashionable, too, to produce small spring driven timepieces with cases based on the larger models. Some were so designed that alarm mechanism could be added and the models appeared in their makers' catalogues with the remark "alarm–cents extra".

These small clocks have recently become collectors' items and form an attractive decoration for a country cottage once they have been restored. The term "Cottage clock" is generally applied to them, but in the makers' catalogues the name "Cottage" was usually reserved for a type that had a rectangular top with a square dial. The glass tablets hiding the pendulum on this type were very small and usually had a conventional design or a small mirror–sometimes even painted bunches of flowers.

Below *A cottage clock by the Seth Thomas Co. and* (Right) *A typical Black Forest product, late 19th century.*

This humble type of clock is not without a certain technical interest. Some manufacturers used circular plates for the movement and other unconventional shapes. Winding holes were placed in various positions on the dial. When movements are found belonging to the type that can be issued either as a timepiece or alarm, those of the timepieces have an unbalanced appearance on account of the empty spaces between the placts where the alarm work should go. The Waterbury Company solved this problem with a separate alarm movement that could be fitted in the bottom of the case, leaving the going part undisturbed.

The E. N. Welch Company brought out a series with a normal escapement, then changed over to a patent escapement.

When the German industry progressed from small workshops to factory production, they copied these clocks and the Junghans Company even sold them at a loss, as they helped to sell other types of clock.

Makers in both countries used to put a paper in the back of the case as on the OG clocks, but many of these have been removed in course of time. It is fascinating trying to decide whether a small clock of this type is German or American. Typical of the German clocks are the pendulums marked RA, with a door fastening that consists of a brass hook set on the side of the case. On many German clocks, the case itself has been stained and polished instead of being veneered.

Small cottage alarm clock by the Waterbury Clock Co., Connecticut. The case is the "sharp Gothic" style.

English carriage clock by James McCabe, Royal Exchange, London, mid-19th century. The plainness of this clock is typical of English work.

Dial of French carriage clock, enameled and in the oriental style. The small dial below is for setting the alarm.

The Carriage Clock

IN the past, traveling was not undertaken unless there was a very good reason for it. Roads were bad, conveyances expensive and the danger of highwaymen provided a further deterrent to would be travelers.

Despite these hazards, in the 17th century a type of watch was developing which was known as a carriage watch. It usually struck the hours, had an alarm and was conveniently made to hang inside a carriage. Various London makers also made small lantern clocks with alarms, that could be packed in boxes together with their weights and hung on the wall of an inn before going to bed, in order to wake their owner the following morning.

The carriage clock we know today really dates from the first half of the 19th century. Roads were improving, and after the spread of railways in the 1840s, traveling was undertaken more extensively than before. The carriage clock was not intended to be hung in a carriage but was a traveling clock that could be used in an hotel bedroom.

The earliest English examples date from 1820–30 and were provided with fusees. The shape was rectangular and glass was usually provided on all four sides revealing the high finish of the movement. A further glass top panel allowed a view of the escapement, which could be a cylinder, lever or duplex.

English examples are rare, for it is France that has specialized in this type of clock. There were two main centers of production: at St Nicholas d'Aliermont, near Dieppe, the rough movements were produced and either finished locally or sent on to Paris; the platforms with the escapements were made in the French horological center of Besançon, and were sent to Paris or to St Nicholas to be added to the rest of the movement.

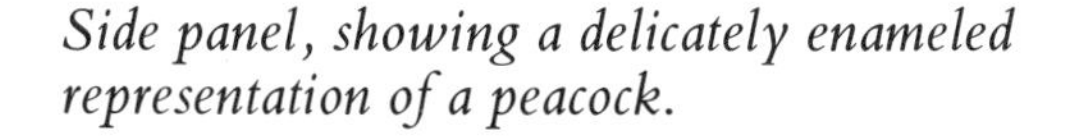

Side panel, showing a delicately enameled representation of a peacock.

Opposite side panel, carrying a cockerel design.

While the case of a carriage clock usually conformed to the rectangular shape, the decoration applied varied enormously and sometimes the glass side panels were replaced by colored enamel plaques. The color of the metal and the design of the handle, too, varied from clock to clock.

Like the cases, the movements varied somewhat. The plainest were simple timepieces, one variation had an alarm—even striking work could be fitted. Clocks could be arranged to repeat the hours and quarters or Grande Sonnerie striking could be provided, that is, the clock struck every quarter indicating the quarter and the previous hour.

For traveling, the carriage clock needed to be small, but some examples up to 8 inches high have been noted. The average height was about 4 inches including the handle. A leather case was always provided, well furnished with cushioning for extra protection. A slide could be moved to reveal the dial if the time was required during the journey.

The popularity of these clocks led to imitation, particularly in America, and such companies as Waterbury and Ansonia put their version of the carriage clock on the market. The finish was naturally not as high as on the French examples.

The carriage clock is extremely popular today and prices are continually rising. It is interesting to see the catalogue of a London firm who imported the clocks about 1910, which reads: "All these clocks are examined and adjusted by our own workmen". For an ordinary timepiece the price was one guinea with cylinder escapement and two guineas with lever escapement. The same clock striking hours and half hours on a gong was £3 and with repeat £3 15s. A variety of other models was offered, the most expensive was only £11 5s.

Right *Box label of a Waterbury watch. An illustration of the factory where the product was made was a traditional feature.*

Below *Waterbury watch with the box in which it was sold. The dial is legible and has been modeled on the American fullplate watch.*

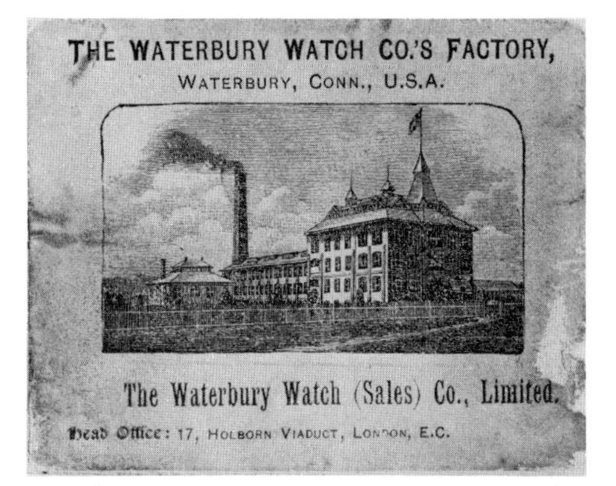

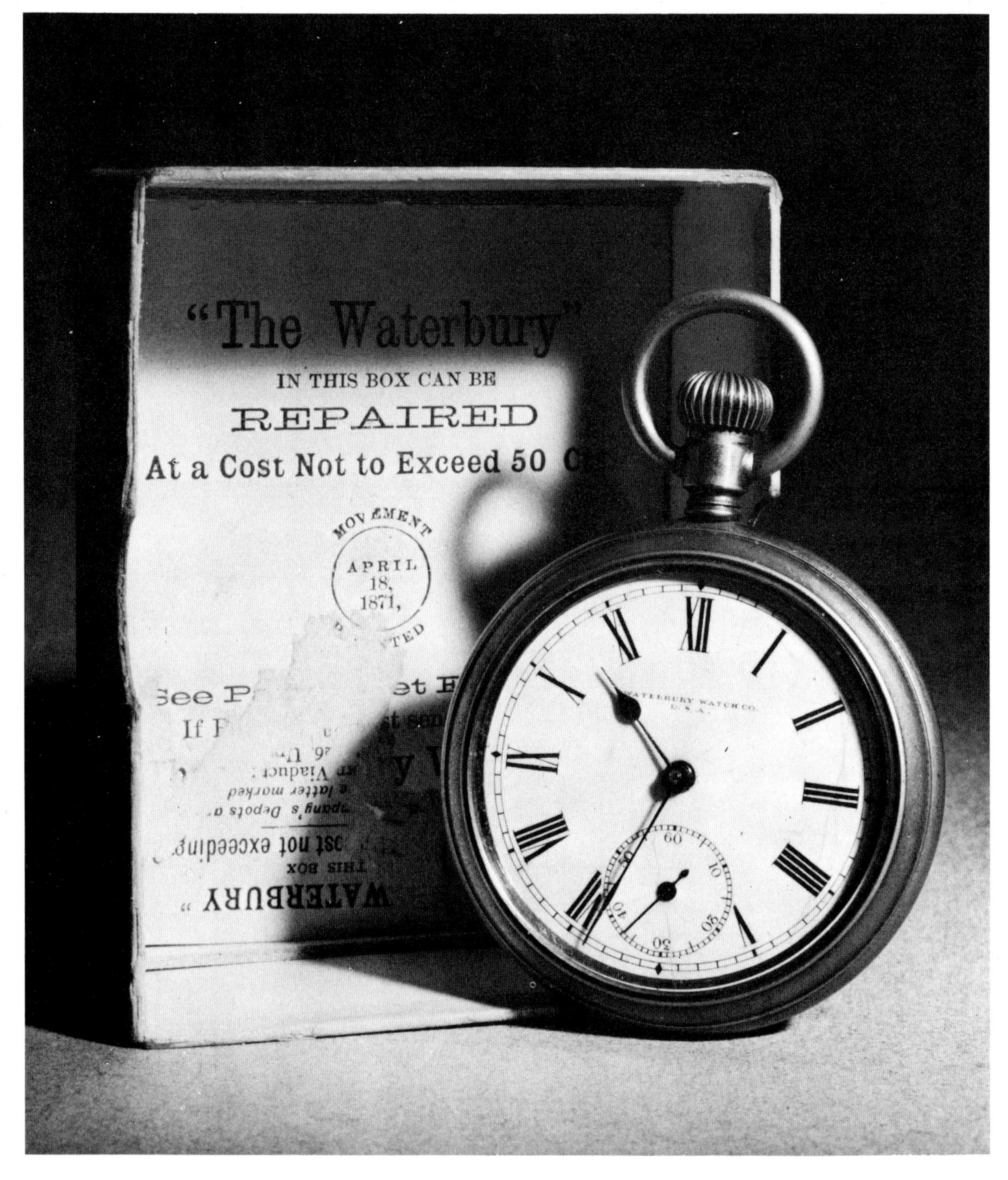

The Waterbury Watch

THE making of watches by machinery in the USA began during the 1850s, but took several years to establish itself. By the 1870s, American machine-made watches were accepted by the public but were all expensive to buy. In 1875, Jason R. Hopkins perfected a model of a watch he claimed could be made for 50 cents, but he did not get financial backing until 1879. The watch then produced was known as the Auburndale Rotary. The company failed in 1883.

In the meantime, Edward A. Locke of Boston, who had been associated with Hopkins in the early part of his venture, came across a watchmaker in Worcester, Massachusetts named D. A. A. Buck, who had exhibited a tiny steam engine at the Centennial Exhibition in Philadelphia in 1876. Locke suggested that Buck might design a cheap watch for him, and at the second attempt a successful model was produced. It had only 58 parts–about half the number of a normal watch–and the mainspring was 9 feet long, occupying the whole of the back of the movement. The movement rotated once per hour and carried the minute hand with it. This idea was virtually the *Tourbillon* of Breguet and meant that better timekeeping would result, as the errors of the watch when going in different positions cancelled each other out.

The escapement used was the "Duplex", which had been used on the better watches in England in the early 19th century, but instead of the complicated scapewheel used in English watches, the wheel in the Buck watch was cut from a brass sheet with alternate long and short teeth, the short ones bent upwards to give the equivalent of two wheels mounted on the same arbor.

The Patents were taken out in 1878 and by 1880 the factory was ready to start production. The company making the watch was the Benedict and Burnham Manufacturing Co. of Waterbury, Connecticut, so the watch became known as the Waterbury. The factory is shown on the box label opposite.

The long mainspring proved tiresome. It took 140 half turns of the button to wind it and some people would hold the button against a wall and then run along the wall to wind up the watch. A music hall joke of the time referred to Waterbury as "the land of eternal spring". The watches bore the warning not to remove the cover of the spring "unless you are a practical watchmaker".

Waterbury watch movement. This is a later model with a short wind. The long and short teeth of the scapewheel can be seen.

The first watches were known as model A and subsequent developments brought in model B, model C and so on. The original price was $4.00, but it later dropped to $2.50. By 1888, production had reached half a million per year. The total series of models ran from A to W and all of these up to series E had rotating movements. A new series was then started which eliminated the long winding arrangements. From series R the watches could be bought with silver or rolled gold cases. Some watches carried the name Charles Benedict instead of Waterbury. In 1898 the name was changed to the New England Watch Company, but the firm was eventually bought out by Ingersoll in 1914.

Testimonials were not lacking. The director of the observatory at Rochester, New York, the President of Columbia College, New York, and Thomas A. Edison all praised the watch, the latter saying that it was the only watch to keep reasonable time in the vicinity of powerful dynamos. During the 1930s, the late Commander R. T. Gould, famous for his work on the Harrison chronometers, carried a Waterbury while doing his investigations into the Loch Ness monster.

The Waterbury project eventually failed because the watch fell into contempt through being given away as an inducement to buy suits of clothes and other articles. The public did not think it worth buying the watch itself. Now, they are collectors' items and change hands for many times their original price. The late C. A. Ilbert himself included them in his collection.

Skeleton clock fitted with striking. The fusee, barrel and wire gong at the rear can be seen here.

Skeleton clock with a large main wheel to give it a longer period of running between windings.

The Skeleton Clock

THE skeleton clock is thought to have had its origin in the Gothic wall clocks of the Middle Ages. These early clocks were made with their movements exposed, the motive being not to display fine finish of parts, but to render visible what was then considered a very wonderful object.

In the late 18th century, when France was producing highly finished clock movements, the idea of using the movement of the clock as a decorative object was revived. The plates were made as narrow as possible to allow the wheels to be seen, and every surface was given a high polish.

The standard English type of movement also began to be produced in skeleton form about 1830, and was a development of the English dial timepiece. While French makers did not often use a fusee, English clocks were almost always fitted with it, and in the skeleton clock it added to the interest of the movement.

The tradition has grown up that these clocks were made by apprentices as their masterpiece when they had completed their training. While certain clocks of this type may have been made by apprentices, there were a number of firms producing them commercially in sufficient number to nullify the general tradition.

The English skeleton clock was always a solid piece of engineering with very elaborate plates incorporating designs of foliage or architectural elements, on which the Gothic motif was often apparent. The basic type was a timepiece only, but an arrangement to strike one blow at the hour, operated by the timekeeping mechanism, was often included. Certain skeleton clocks were provided with striking or chiming, but the addition of further mechanism has led to visual complexity and some of the attraction of the piece has been lost. Skeleton clocks have created sufficient interest for some collectors to specialize in the type, and striking models are now especially popular.

These clocks have often been used as a means of demonstrating a new escapement or horological idea and consequently unconventional examples are sometimes to be found.

The glass shades which should cover these clocks are not always present, but in Victorian times replacements could be easily obtained, for at that time drawing rooms contained numerous objects such as china ornaments or wax fruit under glass shades and the glass works found it worth while to keep them in production.

Striking skeleton clock with an unusual frame design. The bell is not of the hemispherical shape usually found on these clocks.

Typical monumental chime clock, often known as a "Directors'" or "Board room" clock. The name at the bottom is that of the vendor.

The Monumental Chime Clock

THE sounding of chimes at the quarters was known as early as 1390, when the great turret clocks of Wells and Rouen were made. This was also a feature on some of the Gothic wall clocks and on spring driven table clocks of the Renaissance. It was used by London makers from the later 17th century, both in long case and spring driven clocks, and was still being used at the beginning of Queen Victoria's reign.

In the mid-19th century, two developments had an effect on the history of the chiming clock. The new clock for the Houses of Parliament was provided with quarter chimes and immediately won the affection of the general public. Secondly factory production of clocks made it possible to make chiming clocks for a lower price than before.

The term "cheap" as far as English work was concerned, is only relative. The clocks produced were solid, highly finished pieces of machinery with fusees, very thick plates and solid cases. The cases were decorated with carving, applied bronze and other ornamentation inspired by the Great Exhibition of 1851, and were intended for a room in a Duke's mansion or a Board Room. English makers had up to that time always used bells for their chimes, but many of these monumental clocks now had gongs for the purpose of striking the hour. Not only did the gongs take up less space, but the deep tone gave the suggestion that the clock was closely related to Big Ben. Some of the clocks would have looked overpowering even on the large mantelpieces of the time, and so they were often set on brackets at a greater height than the average mantelshelf.

Like many other clocks, the monumental chimes had their imitators, in particular the company of Winterhalder & Hoffmeier of Neustadt, Black Forest. This company had begun by making clocks with wooden frames and brass wheels, gradually changing over to an all metal movement. They produced chiming clocks with fusees and even made their own version of the English dial clock. Cases for the chiming clocks were produced similar to the English ones. Although the company was imitating the English product, their work cannot be disparaged because it was of extremely high quality, and even today their clocks are seen in sale catalogues described as English. Their work was much respected by the English horological trade. The clocks are only identifiable by the letters W&H SCH stamped on the bottom of the movement.

A catalogue of the Edwardian era shows both types of chiming clock, English and foreign. In outward appearance they are very similar, but the imported clocks are sold at about half the price of the English ones.

The monumental chime also had its imitators in America, but the American factories nearly always chose a long case clock as their product and made the cases as large and elaborate as possible. They usually had a glass door to the trunk to reveal the pendulum and weights and the chimes were produced by visible tubular bells.

The American Drop Dial

ENGLISH dial clocks had begun to be made with striking work in the period 1820–40 and were provided with cases having a trunk for a long pendulum. The cases were often decorated with inlay and made of mahogany with a pleasing reddish color. The American factories began to reproduce this model some 30 years later and production in many instances lasted until well into the 20th century. The brass inlay gave way to wood inlay, then to plain cases, sometimes with gold leaf applied to the edge of the window in the trunk. The basic design underwent little change, except that the window in the trunk became larger and the pendulum rod was made of wood and fitted with a larger bob than previously.

It is noteworthy that in the period when the German clock industry was changing over from small workshops to factory production in the 1860–75 period, copies of the English drop dial were produced, some of them of high quality and even containing fusees. If the case of a clock has been lined with blue paper, this may indicate German manufacture. The American factories usually put a label on the case showing the maker's name and sometimes stamped the name on the plates of the movement. The American product always had the type of movement associated with America: open springs, open plates and lantern pinions. In the USA, the clocks are usually known as "schoolhouse clocks", but they were found in other large rooms, from village post offices to western saloons.

Some models were made with additional calendar work and others had the word "Regulator" painted on their glass door, which was meant to inspire confidence in the clock, although a dead beat escapement and compensated pendulum were not necessarily present in the movement. Some of the British prototypes had long pendulums reaching almost to the bottom of the case, but the American clocks generally had shorter pendulums. The reason for this was that the movements were suitable for several different types of case and it would not have been economic to bring out a separate design of movement for each type of case manufactured. Most of the American factories produced the drop dial. Those most frequently sent to Britain were from Ansonia, New York, and from Seth Thomas of Thomaston, Connecticut.

During the years when the American striking dials were popular, the British prototype

had almost ceased to be made. The English dial as a timepiece was in full production, but the trunk type of case was much rarer. Possibly the flood of American clocks discouraged British manufacturers from competing. Certainly, the English striking movement was expensive to produce and the American product comparatively cheap. In the past, most of the clocks of the American drop dial type were used in kitchens, and it is only today, as collectors begin to turn their attention to them, that these clocks are now much sought-after.

Above *Typical American drop dial movement. This is virtually a development of the OG movement provided with spring drive.*

Opposite *American drop dial. The case is apparently inlaid with brass, an effect achieved using suitable colored white metal.*

Below *Three crystal regulators. The clock in the center is plain and of normal height (10 inches). The other two are smaller with decorated cases.*

Bottom *Another view of the three clocks, showing the use of Arabian figures and dial decorations.*

The Crystal Regulator

AT the beginning of the 19th century, clocks in France became very plain, in contrast to the elaborate detail of the previous century. Certain of the leading Paris makers produced clocks of high precision with dead beat escapements and pendulums compensated for changes in temperature. Their cases were simple and rectangular, made of wood and glazed on all four sides–in fact the shape of the clock resembled that of a carriage clock.

In the second half of the 19th century this design was revived in a modified form, and resembled the carriage clock even more, for the case was now made of brass. The well known French circular movement was used, but a dead beat escapement was fitted, and the compensated pendulum, instead of being of the gridiron type with rods of brass and steel, now consisted of two small glass jars of mercury, thus greatly enhancing the appearance of the clock. The Brocot pin pallet escapement was popular and was fitted in some models between the plates, while in others in front of the dial. Although the clocks set out to be precision timekeepers, they still possessed striking, and in some instances more decoration was applied to the cases and dials than would be expected on a precision clock.

The chief disadvantage of the type was its lightness which could lead to accidents if the clock was inadvertantly knocked. A precision clock demands a very firm base and no disturbance, but the brass case of this type may tarnish and need polishing and the glass sides would show the dirt, so the clock would have to be disturbed for purposes of cleaning.

Dials for these clocks were made in numerous designs with both Arabic and Roman figures, but the dial that suited the type best was of white enamel with small Roman figures, on which the winding holes were protected by solid brass rings.

The idea of setting jars of mercury on the pendulum was soon adopted by German and American factories for much cheaper clocks that did not warrant compensated pendulums. The pendulums fitted to these cheaper clocks did not possess real glass jars of mercury but highly polished metal cylinders giving the effect of mercury at a distance. They were, of course, not compensated pendulums at all, but were merely fitted to impress customers.

American factories also produced their version of the French precision clock. The Ansonia Company in particular issued a whole catalogue devoted to variations on the type. The movement designed to be fitted to marble clocks in imitation of the French styles was discarded for one with circular plates, for as the movements in the precision clocks were entirely visible, one of the conventional American type would not suffice. The Ansonia clocks were mostly decorated with bronze castings, but one or two models in the original style were offered. The Waterbury Company also produced clocks to compete with the French ones and named their models after French towns and provinces such as Calais, Dijon and Brittany. A further variation produced in France combined the mercury pendulum with a marble case, with a window through which the pendulum was visible.

The Roskopf Watch

WHILE watches were being made in large numbers in the mid 19th century, they were still undergoing a number of finishing processes that made them expensive. The gathering momentum of industry made it more than ever necessary that employees should be provided with a timepiece, but most of them were too poorly paid to be able to afford a watch.

Georges Frédéric Roskopf (1813–89), of La Chaux de Fonds, had the idea of producing a watch that was robust and accurate and yet would sell for a price as low as 20 francs. He began working on the idea in 1860, and as a beginning designed a simple movement with no superflous adornment, and a case to match. He suppressed the center wheel of the conventional watch, thereby gaining two advantages: there was one part fewer to make, and the spring barrel could be made larger, giving more driving force. The barrel arbor on the Roskopf watch drove the hands through the motion work behind the dial–a similar idea to that used in the old Gothic wall clocks and in the English thirty hour weight movements.

Roskopf also simplified the escapement, providing a lever with pins instead of the conventional shaped pallets, and made the escapement as a separate unit. Keyless work was provided and operated only in the forward direction, but the hands had to be set by the fingers, as on a clock. The stopwork so popular on Swiss watches (fitted to ensure that only the middle turns of the spring were used) was omitted to save expense.

Roskopf watch of the later series, allowing the hands to be altered by the winding button. Note the tail on the minute hand.

Roskopf had at first thought of using a paper dial, but he was worried about the reaction of chemicals in the paper with the metal movement. The case was an alloy of copper, zinc and nickel and made without a joint at the back.

The method of making watches in Switzerland was this: a factory would supply the *ébauche*, or rough movement, then other companies would finish the *ébauche* according to their own style. Having designed his *ébauche*, Roskopf had some difficulty in finding a company to make it and craftsmen to finish it. He encountered much opposition because the workmen did not want to undertake something they were not used to. For this reason he even had to send some of the *ébauches* to Doubs, France, to have them finished.

His efforts were crowned with success in 1865–7 and in 1868 he exhibited the watch at the Paris Exhibition, where to his surprise it won a bronze medal. The next step was to produce a watch with a hand setting mechanism, and he was able to arrange for the *ébauches* to be made in 1870, but the price of the watch had to be increased to 25 francs.

Roskopf took out Patents in several countries but not in Switzerland, so he had many imitators. His work proved to be of great benefit to the Swiss industry, and any cheap watch with a pin pallet escapement is nowadays known as a Roskopf. It should not be forgotten that, in spite of his aim to produce a cheap watch, Roskopf himself insisted on high quality materials and good pay for the workmen, so that the quality of his product remained consistent.

Roskopf watch movement. In the earliest model the piece of metal holding the winding wheels was L-shaped.

Locomotive or marine timepiece from Ansonia catalogue. The octagonal surround to the dial was very popular for this type of clock.

Mobile Clocks

THE carriage clock was essentially a timekeeper for the exclusive use of a traveler, who did not generally consult the clock on the journey, but used it in his hotel room after his day's traveling. Other types of clock have been devised to be used during a journey, and it is these that come under the heading of "mobiles".

Among the earliest were the small clocks handed to the guards of mail coaches to ensure that the coach kept time on its journey. These consisted of small rectangular boxes faced with brass, with an aperture for the dial. The mechanism was locked in so that there was no chance of it being tampered with. The movements resembled watch movements and had a high standard of timekeeping.

The growth of public transport created a demand for accurate timekeeping, but it was in America that the mobile type of clock really came into its own. The Jerome Company's catalogue of 1852 already listed three varieties: 6 inch, 8 inch and 10 inch dials for thirty hour running, and an eight day clock with a 9 inch dial. It was only during the previous decade that American factories had solved the problem of making steel mainsprings cheaply, and yet at this early date clocks were being made with the lever escapement which needed balance springs as well. The making of balance springs required great skill, and yet the American manufacturers were taking the production of these in their stride. The usual case for a mobile clock in the USA was an octagonal wooden surround for the dial. The clocks were advertised for use on locomotives, railway cars and, particularly before the 1860s, on steamboats.

These clocks were not of high quality, having only movements typical of the factories of the time, and it is not surprising that in the 1890s railroads insisted on watches of high quality for running their traffic, so the mobile clocks went out of favor.

Most of the American clock manufacturers seem to have made mobile clocks, supplying a range of about half a dozen different sizes. These clocks must have been popular in the home, too, for some models included strike, alarm and even calendar work.

A specialist form of the mobile was the ship's clock. This was very similar except that it possessed a brass case, and some were arranged to strike the time in ships' "bells", although they did not necessarily allow for dog watches. Ship's clocks were well known in

Mobile clock paper by E. Ingraham & Co., successors to Brewster and Ingrahams, 1870–80.

Pendulum-controlled clock by Brewster and Ingrahams, 1844–52. This provided the inspiration for the design of the mobiles.

England, but there they possessed a movement resembling that of a carriage clock, being of a higher quality than the American product, and precautions were taken to keep the damp out when they were to be used in engine rooms.

A new feature of late 19th century life which had a certain affinity with the ship's engine room, although not mobile, was the power station. Here were found the hot damp conditions of the engine room together with powerful electrical fields that might magnetize a clock and cause it to perform erratically. Special anti-magnetic versions of the ships' clocks were therefore made.

The Eight-day Watch

IN view of the number of special features that have been added to watches in the past, it is rather strange that little attention has been given to lengthening the period of going between windings. The main difficulty in the past was the use of the fusee, which took up a large amount of room. Another difficulty was that a watch to run for eight days needed a large mainspring, which, if used in connection with a fusee, made the watch uncomfortably large. Not until the going barrel could be applied was it possible to make a watch which would run for more than one day on a winding.

One of the pioneers was Robert Westwood of Princes Street, Soho, who took out a patent for an eight day watch in 1829 (No 5850). In Westwood's watch the barrel occupied two thirds of the diameter of the movement and drove an extra wheel in the train. By arranging two extra wheels, the barrel diameter could be made three-quarters of the movement diameter. Westwood claimed that his watch functioned as well as a thirty hour one because of its driving power.

In order to keep the diameter of the watch down, he arranged the movement in two levels, a feature which can be found on some very small wrist watches of today. However, the thickness had to be increased. The Duke of Sussex purchased one of these watches and Westwood afterwards described himself as watchmaker to HRH the Duke of Sussex, but when the Duke died in 1843, only one Westwood watch was included in the sale catalogue of his effects–a gold eight day watch Number 50. This may mean that it was the 50th eight day watch made by Westwood, but this is difficult to say, for comparable specimens today are very rare indeed, as are any watches or clocks by Westwood. (Westwood himself came to an unfortunate end, being murdered by a burglar in 1839.)

It appears that no other eight day watches were made until the turn of the century, when Switzerland produced a model with the escapement on the front of the watch which was visible through the glass, with the dial smaller than usual. The mainspring barrel in these watches occupied the whole diameter of the case and was reminiscent of the early Waterbury watches. Like many other horological pieces they had their imitators. A watch with an ordinary Roskopf type movement was arranged with its balance visible through the glass, and was marked *Façon 8 jours* (8 day style).

These watches are once more becoming popular, for they are being made in Switzerland for a German firm and are sold as replicas. The new models run for eight days, and have 15 jewels, while one model is even provided with a hunting case.

Above *Eight day imitation watch* Façon 8 jours. *This watch is fitted with a normal type of cheap movement and the only connection with the eight day type is the balance visible on the dial side.*

Opposite *Eight day watch* Hebdomas 8 jours. *In this model the mainspring barrel occupies the whole diameter of the movement.*

R
A
8 JOURS
HEBDOMAS

Early 400 day clock of German origin, 1885. It is regulated by moving the two weights on the circular balance nearer to or away from the center.

The 400-day Clock

CLOCKS made to run for a year between windings were known in the late 17th century, but they suffered from the disadvantage of requiring a very heavy driving weight, while in contrast the wheels at the end of the train and the escapement had to be very light.

Aaron D. Crane of New Jersey patented a torsion pendulum arrangement that required far less power. (A torsion pendulum consists of a heavy weight supported by one or more threads, which rotates first in one direction, then in the other, with the thread(s) twisting and untwisting as the pendulum operates. When the thread is twisted, the bob is raised slightly and gravity tends to restore the status quo.) In addition he patented striking work to run for a year. Records of his patent of 1829 have been lost, but his patent of 1841 has been preserved. The Dutch scientist Huygens had experimented with a torsion pendulum in the mid 17th century, but it was arranged to beat comparatively rapidly whereas Crane's vibrated very slowly. Crane's clocks were manufactured in New York and were very much smaller than the 17th century year clocks.

The 400 day clock type really began to be popular from about 1880 onwards when various European firms began making small clocks to be displayed under glass shades with a torsion pendulum and a going period of more than a year—some were even provided with striking. The bob of the pendulum consisted either of a brass disc with two small weights that could be moved in or out for regulation purposes, like the weights on a foliot, or of three brass balls which could be adjusted by a central screw. Some of the clocks were made in France, but they were especially popular with German factories.

In 1904, an American importer copyrighted the name "Anniversary" for these clocks, the idea being that the clock was wound on the anniversary of a wedding, birthday, etc.

After 1945, many German factories began production of various models, and novelties were introduced such as the phases of the moon, coloured pillars, levelling screws for the base and a small cup to prevent the pendulum from being violently disturbed if the clock was moved. Most of these factories have now discontinued production, and only about four companies still make the clocks.

The 400 day clock was sometimes difficult to adjust, but provided it was clean and properly levelled it performed satisfactorily. The type had a reputation for bad timekeeping, but this has arisen because in many instances the clock was not corrected as frequently as one that had to be wound more often. In 1951, the first nickel steel suspensions allowing for temperature compensation became available.

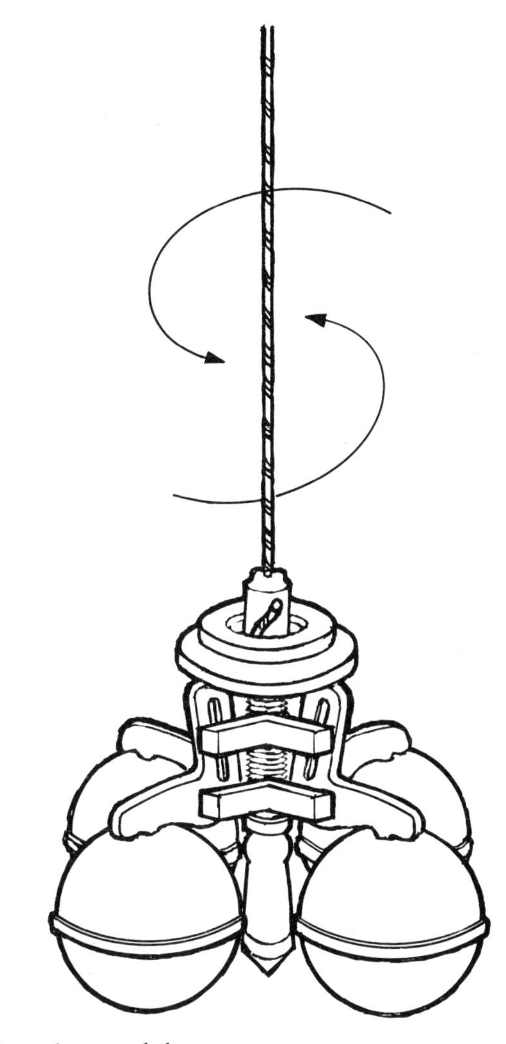

A torsion pendulum.

The torsion pendulum was also applied to the "Atmos" clock, a type driven by variations in temperature. This type of mechanism is particularly suitable for the clock because of the small amount of power it needed to keep it going.

The 400 day clocks are hardly collectors' items yet, but in a few years time it may well be that collectors will try to obtain a specimen from every factory that ever produced them.

Columbus model, 1893. The chrome case is inscribed at the top "R. H. Ingersoll & Bro. New York".

Rear view, including a portrait of Columbus and an early ship.

The Ingersoll Watch

THE Ingersoll brothers, Robert and Charles, were originally in business in New York selling rubber stamps made to order. When the prices in this business dropped, they turned to other lines and became general merchants for metal goods. In 1892 they issued their first mail order catalogue, which included an unbranded watch at $3.95. This watch was the first they sold, but the first true Ingersoll watch was the "Universal", introduced in 1892 and sold wholesale. The "Universals" were manufactured by the Waterbury Clock Co. (not the Waterbury Watch Co.), purchased at 85 cents, sold to dealers at $12.00 a dozen and retailed at $1.50.

The watch was very large,–being about $2\frac{1}{2}$ inches in diameter–and $1\frac{3}{8}$ inches thick. It had only a paper dial, but a seconds hand was included. The movement resembled that of the two-inch timepieces which were being sold in a variety of cases for use in the home.

The great exhibition of 1893 led to the production of a slightly smaller model sold in plain cases, and also a "Columbus" case especially for the occasion.

The Waterbury Clock Co. had been selling similar watches under their own name, but they arranged in 1895 for the Ingersolls to take a batch of 500,000 with a sole agency. 485,000 of these were sold during the year. 1896 saw the arrival of the first watch priced at $1.00—"The watch that made the dollar famous". Although it apparently possessed a winding button, it was in fact wound from the back by a key that folded down like that of an alarm clock. The name of the model was "Yankee". Several other models were introduced at slightly higher prices and in 1898 the firm sold a million watches.

In 1899 an attempt was made to enter the British market and it was planned to sell a million watches in stores over a period of three years. A special Boer War model was included and 200,000 of these were sold at 10s. 6d. each. In 1904, Ingersoll decided to open their own branch in London, and the model they put on the market was called "Crown" as it sold for five shillings. The "Yankee" was not sold on account of the name and also because it was a rear winder.

In 1911, the London branch began assembling parts made by the Waterbury Clock Co.

Earliest Ingersoll movement, virtually that of a small clock with a winding key that folded out of the way.

The Ingersolls had bought the factory of the Trenton Watch Company of New Jersey in 1908 and bought the factory of the New England Watch Company (formerly the Waterbury Watch Company) in 1914. The Ingersoll interest was bought out by the Waterbury Clock Company in 1922.

The very early Ingersoll models are now collectors' items. Although roughly made, the mainspring is contained in a barrel and the teeth of the wheels are quite well cut. The door at the rear is hinged instead of being a snap-on fit and the movement bears the stamp of the Waterbury Clock Company. Later watches have the name Robert H. Ingersoll and Bro. and carry a guarantee paper.

The Railroad Watch

As soon as the railroad system began to develop, the need for accurate timekeeping became an important factor in running the business. Before the introduction of the Block system of signaling, that is, maintaining a certain distance between each train, trains were dispatched at timed intervals, and if a train were delayed, the following one could easily run into it, with disastrous results.

One of the first problems that confronted the new railroads was the observance of local time throughout the country. For instance, Coventry is 6 minutes slow of Greenwich and Birmingham 7 minutes 36 seconds slow, so a train going from Coventry to Birmingham would apparently take 1 minute 36 seconds less on the journey if the time shown by the station clocks were adhered to. The more lines that were opened, the more acute did the problem become, so the London and North Western Railway decided to standardize Greenwich time over the whole system in 1847, and other lines followed suit.

Three railroad watches. A Waltham (top) *of the South Eastern Railway, England. A watch for the same line made by J. W. Benson, London* (center). *A Somerset and Dorset Joint Railway watch made by Seth Thomas, Connecticut.*

In order to maintain good timekeeping, the operational staff had to be issued with watches, and these watches had to be of a high standard of accuracy, capable of standing up to the hard knocks of railroad life. Watches used by engine men would in addition have to be resistant to dirt and damp. The earliest railroad watches had lever escapements and fusees, being direct descendants of the traveling clocks issued to guards on mail coaches. With the spread of factory-made watches and the abandonment of the fusee, railroads found that sturdy accurate watches could be obtained more cheaply that would do the job just as effectively. Some lines, such as the London Brighton and South Coast and the Somerset and Dorset, went in for American watches, while others adhered to English makers but purchased watches with going barrels. Keyless watches appeared towards the end of the 19th century, but many old keywinders had a long life until the grouping of railroads in 1923.

In America, the running of trains was done on the dispatching system using the telegraph, but the need for accurate timekeeping was just as acute. The USA adopted time zones in 1883, but before 1890 there were no special specifications for a railroad watch.

After a disastrous train wreck in 1891, for which the blame was put on faulty timekeeping, Webb C. Ball, a watchmaker of Cleveland, Ohio, was asked to specify standards for railroad watches. In 1899 the watch inspectors' convention accepted the following stipulation: the watches must keep time in five different positions correct to 30 seconds a week with a temperature variation of 40° to 90°F and be so stamped on the plates. The watches had to be wound at the 12 o'clock position but set by a lever beside the dial so that any accidental rubbing of the button in the wearer's pocket would not alter the hands. They had to have a double roller escapement and bold Arabic figures and hands. A standard mark had to be stamped on the back plate.

Certain manufacturers made watches to Ball's specifications and put his name on the product—manufacturers such as Elgin, Hamilton, Hampden, Howard, Illinois and Waltham. Many US railroads adopted Ball's standards.

Wrist watches began to be accepted for use on US railroads in the 1960s and the last pocket watches were made during the same decade. With the dwindling in number of American watch factories, it is possible that Swiss models will be used on American railroads in future.

Two railway watch movements belonging to the London, Brighton and South Coast railway, supplied by Waltham, USA. The back plate of the Benson movement comes from a watch belonging to the South Eastern railway and is London made.

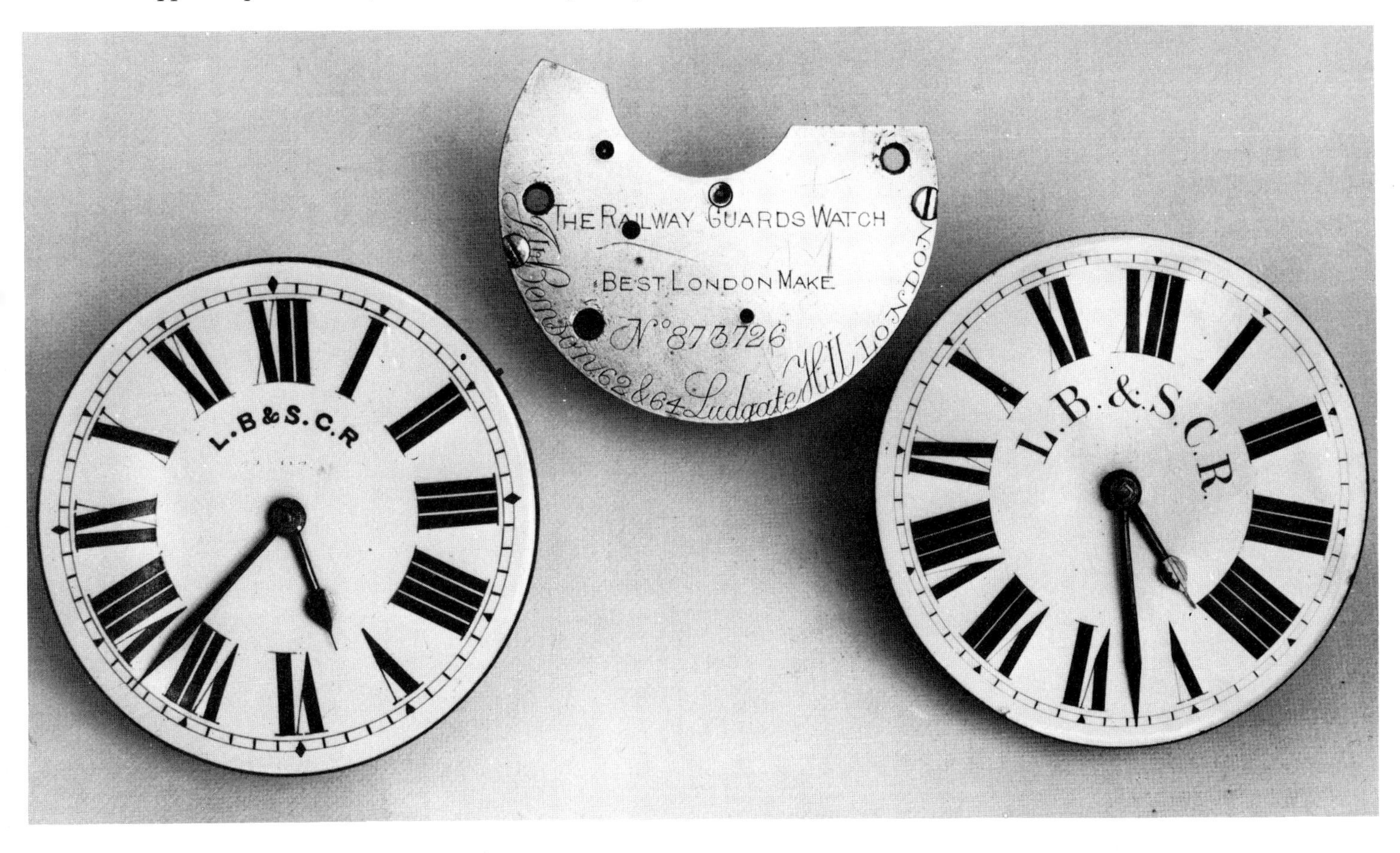

The Military Watch

IN the 18th century, army officers supplied their own watches, and there were no special features to distinguish them from watches used by civilians. After the Napoleonic wars, the Duke of Wellington became interested in the products of Breguet, and other British officers followed his example, so setting the fashion for precision timekeeping in the army. French officers on active service had been using small traveling clocks with repeat and alarm, and these are known today as *pendules d'officier*. They are artistic pieces of work but too ornate for army life.

With the increasing importance of accurate timekeeping during the 19th century, it became necessary to provide specially designed watches for military use. Not only did the watches have to be accurate, but also capable of standing up to the tough conditions. The cases were made as strong as possible which resulted in the watches being slightly larger than those for civilian use. Another important feature was that the front and the back screwed on to give maximum protection from dirt and damp. A seconds hand became a necessity.

It is believed that the first wrist watches on active service were issued to officers of the German Navy in 1911, but the war of 1914–18 demonstrated the usefulness of the wrist watch and showed that it could stand up to exacting conditions and also keep accurate time.

The latest developments in military watches are illustrated by the Doxa "Army". This is a wrist watch with automatic winding, protected against shock and provided with a calendar. The case is oxidized steel and does not give reflections, and by means of a special sealing system is able to withstand water pressure of 30 atmospheres. The button is sunk into the rim of the case for extra protection. The bezel can be rotated, and bears luminous figures to indicate the time before an action begins, while the dial is legible by day and night and special attention has been given to the hands for maximum clarity. The center seconds hand has a luminous tip, making the reading of seconds easy in the dark, and to ensure maximum legibility the glass is made flat and is unscratchable as well as unbreakable.

This watch represents the ultimate in the mechanical watch. Further watch developments will probably be in the field of electronics and will not involve any moving parts at all—a far cry from the erratic performance of the stackfreed and Nuremberg egg.

Below left, *a watch by Elgin, USA. The other two are by H. Williamson, London. The center model has a luminous dial.*

Bottom *The Doxa "Army" military watch. Its appearance is functional–it comes in a camouflaged containing bag.*